Robust Stabilisation and H^∞ Problems

Mathematics and Its Applications

Volume 482

Robust Stabilisation and H^{∞} Problems

by

Vlad Ionescu

Faculty of Automatic Control and Computers,
Bucharest Polytechnical University,
Bucharest, Romania

and

Adrian Stoica

Faculty of Aerospace Engineering,
Bucharest Polytechnical University,
Bucharest, Romania

SPRINGER SCIENCE+BUSINESS MEDIA, B.V.

A C.I.P. Catalogue record for this book is available from the Library of Congress.

ISBN 978-94-010-5978-7 ISBN 978-94-011-4702-6 (eBook)
DOI 10.1007/978-94-011-4702-6

Printed on acid-free paper

Dedicated to the memory of Professor Aristide Halanay

TABLE OF CONTENTS

PREFACE

It is a matter of general consensus that in the last decade the H^∞-optimization for robust control has dominated the research effort in control systems theory. Much attention has been paid equally to the mathematical instrumentation and the computational aspects. There are several excellent monographs that cover the standard topics in the area. Among the recent issues we have to cite here *Linear Robust Control* authored by Green and Limebeer (Prentice Hall 1995), *Robust Controller Design Using Normalized Coprime Factor Plant Descriptions* – by McFarlane and Glover (Springer Verlag 1989), *Robust and Optimal Control* – by Zhou, Doyle and Glover (Prentice Hall 1996). Thus, when the authors of the present monograph decided to start the work they were confronted with a very rich literature on the subject. However two reasons motivated their initiative.

The *first* concerns the theory in which the whole development of the book was embedded. As is well known, there are several ways of approaching H^∞ and robust control theory. Here we mention three relevant directions chronologically ordered: a) the first makes use of a generalization of the Beurling–Lax theorem to Krein spaces; b) the second makes use of a generalization of Nevanlinna–Pick interpolation theory and commutant lifting theorem; c) the third, and probably the most attractive from an elevate engineering viewpoint, is the two Riccati equations based approach which offers a complete solution in state space form. Unlike the above mentioned approaches our option was orientated towards the so called *generalized Popov-Yakubovich theory.* This choice has been strongly motivated by the recent encouraging results obtained by the authors in this field. In fact, a theory which generalizes the classical Popov–Yakubovich *positiveness theory* to the *indefinite sign case* has been developed. The results are easily derived and encapsulated in a comprehensive form.

The *second* reason is of practical origin. Firstly, the book has been conceived such that each theoretical result serves as an authentic support for solving significant tasks of modern control theory. Secondly, several case studies emphasize the efficiency of the proposed methods both in computational aspects and in achieving control design specifications.

The above arguments suggest immediately the audience. Thus a first group of readers to whom the monograph is addressed consists of those researchers involved in advanced control topics and engineers acting in the forefront of applied control theory. A second group, incorporating Ph.D. students, could be interested in the present monograph as well. In fact, the book is available to any reader possessing an appropriate mathematical background. Specialists in numerical computation could be also interested in the content of this book.

The present monograph is divided into six chapters.

In *Chapter 1* some basic concepts and known results in linear systems theory are briefly recalled. Nonorthodox subjects such as Hardy spaces, L^∞-norms, Hankel and Toeplitz operators and balanced forms fully exploited in the whole development are treated as well.

Chapter 2 contains the main theoretical framework on which the present monograph is built, namely the generalized Popov–Yakubovich theory. The theoretical results derived in this chapter are particularly orientated towards the existence conditions of the stabilizing solution of the so called *Kalman–Popov–Yakubovich system in 'J-form'*. Such a mathematical object seems to be the most appropriate tool for a suitable description of the stability and attenuation requirements as they naturally are posed in the control synthesis tasks. The main result emphasizing the relationship between the invertibility condition of a certain Toeplitz operator and the stabilizing solution to the Kalman–Popov–Yakubovich system of indefinite sign is expressed in terms of the so called *signature condition.*

Chapter 3 reveals how the signature condition proves to be a very efficient for finding a complete solution to the regular H^∞control problem. Thus necessary and sufficient solvability conditions are easily derived and explicit formulae for the H^∞ controller are also given.

Chapter 4 deals with various cases of the Nehari problem. State space solutions in terms of the generalized Popov–Yakubovich theory are presented. Furthermore, using a singular perturbations technique, optimal solutions are derived as well. The construction of the optimal solution for the two-block Nehari problem is illustrated by two numerical examples. The optimal solution for the one-block case has been used in a case study concerning the robust design problem with respect to *normalized left coprime factorization* of the plant. Explicit formulae for such solution are derived.

Chapter 5 describes an optimal solution of the H^∞control problem which is also obtained by employing the singular perturbations method. A reduced order stabilizing controller which ensures the best level of attenuation for a regular H^∞ problem is given. Moreover, all ill conditioned computations which usually appear when using the formulae for the suboptimal solution near the optimum, are completely avoided. The construction of the optimal H^∞ controller is illustrated by two case studies. Thus a model matching problem and an optimal robust design with respect to *additive uncertainty* are presented.

In *Chapter 6* we consider two singular H^∞ problems arising from the robust design with respect to *multiplicative uncertainty* and the *sensitivity minimization*, respectively. Necessary and sufficient conditions for solving these problems are obtained through a known *linear matrix inequalities*-based technique. Explicit formulae for the corresponding controllers in terms of the solutions to these inequalities are also derived. Several numerical examples serve as a support in arguing the efficiency of the proposed procedures.

Most of the design examples considered in this monograph concern the aircraft domain. This fact was motivated by the increasing interest in modern design techniques to achieve better aircraft performances. However, the theoretical results described in this book can also be successfully used in other control engineering domains.

Much attention has been paid to the numerical aspects of the design methods proposed in this book. Thus common procedures included in any software package for control design and analysis proved to be sufficient for an accurate numerical implementation.

It is a question of honor to stress that most of the results included in this book were primarily presented in the Seminar on Differential Equations and Dynamic Systems directed by Professor Halanay, who died recently.

Finally, we are indebted to Kluwer's publishing staff for helpful and friendly assistance.

Bucharest 1998

Vlad Ionescu

Adrian Stoica

ACRONYMS, NOTATIONS, AND SYMBOLS

ACRONYMS: RICCATI THEORY

Σ	Popov triplet
Π_Σ	Popov function associated with the Popov triplet Σ
EHP(Σ)	extended Hamiltonian pencil associated with the Popov triplet Σ
ARE(Σ)	algebraic Riccati equation associated with the Popov triplet Σ
KPYS(Σ, J)	Kalman–Popov–Yakubovich system associated with the Popov triplet Σ and the sign matrix J

ACRONYMS: SYSTEMS THEORY AND ROBUST CONTROL

LMI	linear matrix inequality
DEP	disturbance estimation problem
DFP	disturbance feedforward problem
OEP	output estimation problem
GM	gain margin
PM	phase margin
$\mathrm{M}_{\max}$	the largest absolute value of the elements of matrices giving the realization of a system
$\mathrm{S}_{\mathrm{mar}}$	relative internal stability margin
NLCF	normalized left coprime factorization

MISCELLANEOUS

$:=$	the left hand side defined by the expression in the right hand side
$=:$	the right hand side is defined by the expression in the left hand side
$\square$	end of the proof
$x \equiv y$	x is identical to y

NUMBERS AND SETS

$\varepsilon \searrow a$	ε tends to a with values greater than a
$\mathbf{R}$	field of real numbers
$\mathbf{C}$	field of complex numbers
$\mathbf{C}^-$	open left half plane
$\mathbf{C}^+$	open right half plane
$j\mathbf{R}$	imaginary axis

OPERATORS

$\mathcal{R}^*$	adjoint of $\mathcal{R}$
$\gg$	coerciveness, *i.e.*, $\mathcal{R} \gg 0$ if $\mathcal{R}^* = \mathcal{R}$ and $\langle u, \mathcal{R}u \rangle \geq \rho \|u\|^2$ for some $\rho > 0$
$\mathcal{R}^{-1}$	bounded inverse of $\mathcal{R}$ (for boundedly invertible $\mathcal{R}$)
$\|\mathcal{R}\|$	operator norm of $\mathcal{R}$
$\mathrm{H}^c_{\mathcal{T}}(\mathrm{H}^a_{\mathcal{T}})$	causal (anticausal) Hankel operator of $\mathcal{T}$
$\mathrm{T}^c_{\mathcal{T}}(\mathrm{T}^a_{\mathcal{T}})$	causal (anticausal) Toeplitz operator of $\mathcal{T}$
$\|\mathcal{R}\|_H$	Hankel norm of $\mathcal{R}$
$\Psi^c(\Psi^a)$	causal (anticausal) controllability operator
$\Theta^c(\Theta^a)$	causal (anticausal) observability operator
$\dot{x}(t)$	derivative of $x(t)$

RATIONAL MATRICES AND SYSTEMS

$\left[\begin{array}{c\|c} A & B \\ \hline C & D \end{array}\right](s)$	$:= C(sI - A)^{-1}B + D$
$G^*(s)$	adjoint ($:= G^T(-s)$)
RL^∞ ($\mathrm{RL}^\infty_{p\times m}$)	set of $(p \times m)$ proper rational matrix functions with no poles on $j\mathbf{R}$ axis
RH^∞_+ ($\mathrm{RH}^\infty_{+,p\times m}$)	set of $(p \times m)$ proper rational matrix functions with all the poles in $\mathbf{C}^-$
RH^∞_- ($\mathrm{RH}^\infty_{-,p\times m}$)	set of $(p \times m)$ proper rational matrix functions with all the poles in $\mathbf{C}^+$
$LFT(T, K)$	linear fractional transformation of T with K
$T \bigotimes K$	Redheffer product of T with K

SPACES OF FUNCTIONS

$\mathrm{L}^{2,m}$	Hilbert space of square norm Lebesgue integrable $\mathbf{R}^m$-valued functions on $\mathbf{R}$
$\mathrm{L}^{2,m}_+$	closed subspace of $\mathrm{L}^{2,m}$ of functions with support in $[0, \infty)$
$\mathrm{L}^{2,m}_-$	closed subspace of $\mathrm{L}^{2,m}$ of functions with support in $(-\infty, 0]$

P_+^r	orthogonal projection of $\mathrm{L}^{2,r}$ onto $\mathrm{L}_+^{2,r}$
P_-^r	orthogonal projection of $\mathrm{L}^{2,r}$ onto $\mathrm{L}_-^{2,r}$
$\langle \cdot, \cdot \rangle$	inner product (the Hilbert space is implicit)

VECTOR SPACES AND MATRICES

$\|x\|$	Euclidian norm of the vector $x \in \mathbf{R}^n$
I_n	$n \times n$ identity matrix
0	zero matrix (not necessarily square)
$\Lambda(A)$	union of the eigenvalues of A
A^T	transpose of A
A^{-1}	inverse of A (for A invertible)
A^*	complex conjugate transpose of A
$A \geq B$	$A - B$ is positive semi-definite (for Hermitian A and B)
$A > B$	$A - B$ is positive-definite (for Hermitian A and B)
$\det A$	determinant of A
$\operatorname{rank} A$	rank of A
$\rho(A)$	spectral radius of A
$\operatorname{sgn} A$	sign matrix of A (for Hermitian A)
$\operatorname{diag}(A_1, ..., A_k)$	diagonal matrix with diagonal elements A_1,...,A_k
$\lambda M - N$	matrix pencil
$\Lambda_f(M, N)$	union of finite generalized eigenvalues of $\lambda M - N$
$\Lambda_f(M, N)\|\mathcal{V}$	restriction of $\Lambda_f(M, N)$ to the deflating subspace $\mathcal{V}$

CHAPTER 1

LINEAR SYSTEMS: SOME PREREQUISITES

The aim of this chapter is to recall some basic notions and results of the linear systems theory. Parts of these results are now common facts while others are not usually encountered in the classical literature on the subject. Thus, besides the notions of evolution, stability, controllability and observability, we shall also refer to some elementary facts concerning Hardy spaces, L^∞-norms, L^2-forced evolution, Hankel and Toeplitz operators, *etc.*. As the chapter has been thought of as an *aide-memoire* all the results are sketched without proofs.

1.1. STATE SPACE DESCRIPTION AND THE INPUT/OUTPUT MAP

Definition 1.1 *Any matrix quadruple* $(A,B,C,D) \in \mathbf{R}^{n\times n} \times \mathbf{R}^{n\times m} \times \mathbf{R}^{p\times n} \times \mathbf{R}^{p\times m}$ *with the mathematical significance given by*

$$\begin{aligned} \dot{x} &= Ax + Bu \\ y &= Cx + Du \end{aligned} \tag{1.1}$$

where $x(t) \in \mathbf{R}^n$, $u(t) \in \mathbf{R}^m$ *and* $y(t) \in \mathbf{R}^p$ *are the* state, *the* input *and the* output, *respectively, will be termed a (continuous, time-invariant)* linear system. □

If $x(0) = \xi$ then the initial value problem associated with the first equation in (1.1) has the solution

$$x(t) = e^{At}\xi + \int_0^t e^{A(t-\tau)}Bu(\tau)d\tau \qquad \forall t \in \mathbf{R} \tag{1.2}$$

where u is any locally Lebesgue integrable $\mathbf{R}^m$-valued function on $\mathbf{R}$.

The first term in (1.2) describes the *free evolution* and the second one describes the *forced evolution.* The corresponding transfer matrix of the system (1.1) is the mapping

$$\mathbf{C} \ni s \mapsto \left[\begin{array}{c|c} A & B \\ \hline C & D \end{array}\right](s) = T(s) = C(sI-A)^{-1}B + D \tag{1.3}$$

defined for all $s \in \mathbf{C} \setminus \Lambda(A)$, where $\Lambda(A)$ is the spectrum of the matrix A. Clearly $T(s)$ is a proper rational matrix function, *i.e.* each entry of $T(s)$, say $t_{ij}(s)$, is a proper rational function, that is, the degree of the numerator does not exceed the degree of the denominator. As is well known,

$$\mathcal{L}^{-1}T(s)\hat{u}(s) = \int_0^t Ce^{A(t-\tau)}Bu(\tau)d\tau, \tag{1.4}$$

provided u is Laplace transformable with the Laplace transform $\hat{u}$ ($\mathcal{L}u = \hat{u}$, $u = \mathcal{L}^{-1}\hat{u}$). Usually instead of (1.1) we shall adopt the 'mixed' representation

$$T = \left[\begin{array}{c|c} A & B \\ \hline C & D \end{array}\right] \tag{1.5}$$

which aggregates simultaneously the input/output behavior and the state space representation that generated it.

Two systems (A, B, C, D) and $\left(\widehat{A}, \widehat{B}, \widehat{C}, \widehat{D}\right)$ are called *equivalent*, and we denote it as $(A, B, C, D) \sim \left(\widehat{A}, \widehat{B}, \widehat{C}, \widehat{D}\right)$, if they are linked via a state space similarity $\hat{x} = Sx$, that is $\widehat{A} = SAS^{-1}$, $\widehat{B} = SB$, $\widehat{C} = CS^{-1}$, $\widehat{D} = D$. Clearly

$$T = \left[\begin{array}{c|c} A & B \\ \hline C & D \end{array}\right] = \left[\begin{array}{c|c} \hat{A} & \hat{B} \\ \hline \hat{C} & \hat{D} \end{array}\right]. \tag{1.6}$$

Let $T_1 = \left[\begin{array}{c|c} A_1 & B_1 \\ \hline C_1 & D_1 \end{array}\right]$ and $T_2 = \left[\begin{array}{c|c} A_2 & B_2 \\ \hline C_2 & D_2 \end{array}\right]$ of dimensions $p_1 \times m_1$ and $p_2 \times m_2$ be given. For $p_1 = p_2$, $m_1 = m_2$ we have

$$T_1 + T_2 = \left[\begin{array}{cc|c} A_1 & 0 & B_1 \\ 0 & A_2 & B_2 \\ \hline C_1 & C_2 & D_1 + D_2 \end{array}\right]. \tag{1.7}$$

For $m_2 = p_1$ we have

$$T_2T_1 = \left[\begin{array}{cc|c} A_1 & 0 & B_1 \\ B_2C_1 & A_2 & B_2D_1 \\ \hline D_2C_1 & C_2 & D_2D_1 \end{array}\right]. \tag{1.8}$$

If D in (1.5) is nonsingular then

$$T^{-1} = \left[\begin{array}{c|c} A - BD^{-1}C & BD^{-1} \\ \hline -D^{-1}C & D^{-1} \end{array}\right]. \tag{1.9}$$

1.2. DICHOTOMY. $\mathbf{RL}^\infty, \mathbf{RH}_+^\infty$ AND $\mathbf{RH}_-^\infty$ SPACES

Definition 1.2 *$A \in \mathbf{R}^{n\times n}$ is called* dichotomous *if $\Lambda(A) \cap j\mathbf{R} = \emptyset$. The number of eigenvalues placed in $\mathbf{C}^- := \{s : Re\, s < 0\}$ is called the* rank of the dichotomy. □

The following result holds:

Theorem 1.1 *A is dichotomous if and only if there exists a projection* Π *of* $\mathbf{R}^n$, $\alpha > 0$ *and* $\rho_0 \geq 1$ *such that*

$$\left\| \Pi e^{At} \right\| \leq \rho_0 e^{-\alpha t} \quad \textit{for} \quad t \geq 0 \tag{1.10}$$

and

$$\left\| (I - \Pi) e^{At} \right\| \leq \rho_0 e^{\alpha t} \quad \textit{for} \quad t \leq 0. \tag{1.11}$$

□

Definition 1.3 *If* $\Pi = I$, *A is called* stable *and if* $\Pi = 0$, *A is called* antistable. □

Let $L^{2,r}$ be the Hilbert space of square norm Lebesgue integrable $\mathbf{R}^r$-valued functions on $\mathbf{R}$, and denote by $\| \ \|_2$ the appropriate norm. Then the following result is of relevant significance.

Theorem 1.2 *Let A in (1.1) be dichotomous. Then for each* $u \in L^{2,m}$ *the first equation in (1.1) has a unique solution x in* $L^{2,n}$ *given explicitly by*

$$x(t) = \int_{-\infty}^{t} \Pi e^{A(t-\tau)} Bu(\tau) d\tau - \int_{t}^{\infty} (I - \Pi) e^{A(t-\tau)} Bu(\tau) d\tau. \tag{1.12}$$

Furthermore, there exists $\rho > 0$ *such that*

$$\|x\|_2 \leq \rho \|u\|_2 \qquad \forall u \in L^{2,m}, \tag{1.13}$$

that is, the first equation (1.1) defines a linear bounded input state operator from $L^{2,m}$ *to* $L^{2,n}$. □

Corollary 1.1 *If A in (1.1) is dichotomous then (1.1) defines a linear bounded input-output operator* $\mathcal{T} : L^{2,m} \to L^{2,p}$.

Proof. Use (1.12) and (1.13) in conjunction with the second equation in (1.1). □

Definition 1.4 *Let* (A, B, C, D) *be a linear system. Then*

(1) $\lambda \in \mathbf{C}$ *is called an* uncontrollable (unobservable) *mode if*

$$\text{rank} \begin{bmatrix} \lambda I - A & B \end{bmatrix} < n \ \left(\text{rank} \begin{bmatrix} \lambda I - A \\ C \end{bmatrix} < n\right)$$

(2) The pair (A, B) $((C, A))$ *is called* controllable (observable) *if the system has no uncontrollable (unobservable) modes. Any system for which the pairs* (A, B) *and* (C, A) *are controllable and observable is called* minimal.

(3) The pair (A, B) $((C, A))$ *is called* stabilizable (detectable) *if the system has all the uncontrollable (unobservable) modes in* $\mathbf{C}^-$. □

Proposition 1.1 *The pair (A, B) ((C, A)) is stabilizable (detectable) if and only if there exists $F \in \mathbf{R}^{m \times n}$ ($K \in \mathbf{R}^{n \times p}$) such that $A + BF$ ($A + KC$) is stable.* □

If (A, B, C, D) is a system with (A, B) and (C, A) stabilizable and detectable, respectively, and F and K are such that $A + BF$ and $A + KC$ are stable, then the system

$$(A + BF + KC + KDF, -K, F, 0)$$

is called the *Kalman compensator* of (A, B, C, D) and it stabilizes the system (A, B, C, D).

Let us introduce the following sets:

$\mathrm{RL}^{\infty}_{p\times m}$ is the set of all $p \times m$ proper rational matrix functions with no poles on $j\mathbf{R}$

$\mathrm{RH}^{\infty}_{+,p\times m}$ is the set of all $p \times m$ proper rational matrix functions that have all the poles placed in $\mathbf{C}^-$

$\mathrm{RH}^{\infty}_{-,p\times m}$ is the set of all $p \times m$ proper rational matrix functions that have all the poles placed in $\mathbf{C}^+ (\mathbf{C}^+ := \{s : Re\, s > 0\})$.

Clearly $\mathrm{RH}^{\infty}_{+,p\times m}$, $\mathrm{RH}^{\infty}_{-,p\times m} \subset \mathrm{RL}^{\infty}_{p\times m}$. $\mathrm{RL}^{\infty}_{p\times m}$ can be organized as a normed linear space. Indeed, for $T \in \mathrm{RL}^{\infty}_{p\times m}$, let

$$\|T\|_{\infty} := \sup_{\omega \in \mathbf{R}} \|T(j\omega)\| = \sup_{\omega \in \mathbf{R}} \overline{\sigma}(T(j\omega)) \tag{1.14}$$

termed as the L^{∞}-norm of T. Here $\overline{\sigma}$ stands for the largest singular value. One can easily check that $\|T\|_{\infty}$ is really a norm. In fact $\mathrm{RL}^{\infty}_{p\times m}$ can be identified with a subspace of the Banach space $\mathrm{L}^{\infty}_{p\times m}(j\mathrm{R})$ and $\mathrm{RH}^{\infty}_{+,p\times m}$ ($\mathrm{RH}^{\infty}_{-,p\times m}$) is a subspace of the Hardy space $\mathrm{H}^{\infty}_{+,p\times m}$ ($\mathrm{H}^{\infty}_{-,p\times m}$).

Theorem 1.3 *Let*

$$T = \left[\begin{array}{c|c} A & B \\ \hline C & D \end{array}\right] \tag{1.15}$$

with (A, B) and (C, A) stabilizable and detectable. Then:

(1) $T \in \mathrm{RL}^{\infty}_{p\times m}$ if and only if A is dichotomous

(2) If $T \in \mathrm{RL}^{\infty}_{p\times m}$ and $\mathcal{T} : \mathrm{L}^{2,m} \to \mathrm{L}^{2,p}$ is the linear bounded input-output operator associated with (1.15) then

$$\|T\|_{\infty} = \|\mathcal{T}\|. \tag{1.16}$$

□

1.3. HANKEL AND TOEPLITZ OPERATORS

Let $\mathrm{L}_+^{2,r}(\mathrm{L}_-^{2,r})$ be the closed subspace of $\mathrm{L}^{2,r}$ whose functions are of positive (negative) support. Clearly $\mathrm{L}^{2,r} = \mathrm{L}_+^{2,r} \oplus \mathrm{L}_-^{2,r}$. Let $P_+^r(P_-^r)$ be the projection of $\mathrm{L}^{2,r}$ onto $\mathrm{L}_+^{2,r}(\mathrm{L}_-^{2,r})$. Let

$$T = \left[\begin{array}{c|c} A & B \\ \hline C & D \end{array}\right] \tag{1.17}$$

be a dichotomous system. Then, as we have seen before, (1.17) defines a linear bounded input–output operator $\mathcal{T} : \mathrm{L}^{2,m} \to \mathrm{L}^{2,p}$.

Definition 1.5 *Let $\mathcal{T}$ be introduced above. Then the following operators*

$$\begin{aligned} \mathbf{H}_{\mathcal{T}}^c &= P_+^p \mathcal{T} P_-^m \\ \mathbf{H}_{\mathcal{T}}^a &= P_-^p \mathcal{T} P_+^m \\ \mathbf{T}_{\mathcal{T}}^c &= P_+^p \mathcal{T} P_+^m \\ \mathbf{T}_{\mathcal{T}}^a &= P_-^p \mathcal{T} P_-^m \end{aligned} \tag{1.18}$$

are termed the causal Hankel operator, anticausal Hankel operator, causal Toeplitz operator, anticausal Toeplitz operator, *respectively associated with* $\mathcal{T}$. □

In order to simplify the notations and, in addition, to stress the 'system origin' of the operators introduced above, the left hand sides in (1.18) will be usually (re-) denoted as $\mathbf{H}_T^c$, $\mathbf{H}_T^a$, $\mathbf{T}_T^c$, $\mathbf{T}_T^a$ where T is the transfer matrix (1.17).

Definition 1.6 *Assume that A in (1.17) is dichotomous. Then we shall say that T is* causal (anticausal) *if $\mathbf{H}_T^a = 0$ ($\mathbf{H}_T^c = 0$).* □

One can easily check that, in the conditions of Definition 1.6, T is causal (anticausal) if and only if $\mathbf{T}_T^c = \mathcal{T} P_+^m$ ($\mathbf{T}_T^a = \mathcal{T} P_-^m$).

Theorem 1.4 *Let T as in (1.17) with A dichotomous. Then T is causal (anticausal) if and only if A is stable (antistable).* □

Definition 1.7 *Let (A, B, C, D) be given and assume that A is stable (antistable). Then the following operators*

$$\Psi^c : \mathrm{L}_-^{2,m} \to \mathbf{R}^n \ \left(\Psi^a : \mathrm{L}_+^{2,m} \to \mathbf{R}^n\right)$$

and

$$\Theta^c : \mathbf{R}^n \to \mathrm{L}_+^{2,p} \ \left(\Theta^a : \mathbf{R}^n \to \mathrm{L}_-^{2,p}\right)$$

made explicit by

$$\Psi^c u = \int_{-\infty}^{0} e^{-At} Bu(t)dt \ \left(\Psi^a u = -\int_{0}^{\infty} e^{-At} Bu(t)dt\right) \tag{1.19}$$

and

$$\left(\Theta^c x\right)(t) = e^{At}x, \ t \geq 0 \ \left(\left(\Theta^a x\right)(t) = e^{At}x, \ t \leq 0\right), \tag{1.20}$$

are called the causal (anticausal) controllability *and* observability operators, *respectively.* □

Theorem 1.5 *Let T as in (1.17) with A stable (antistable). Then*

(1)

$$\mathbf{H}_T^c = \Theta^c \Psi^c \ (\mathbf{H}_T^a = \Theta^a \Psi^a) \tag{1.21}$$

(2)

$$\|T\|_H^c := \|\mathbf{H}_T^c\| = \overline{\sigma}\left(P^c Q^c\right) \ \left(\|T\|_H^a := \|\mathbf{H}_T^a\| = \overline{\sigma}\left(P^a Q^a\right)\right)$$

where $\|T\|_H^c$ ($\|T\|_H^a$) is termed as the causal (anticausal) Hankel norm *and $P^c(P^a)$ and $Q^c(Q^a)$ are the* causal (anticausal) controllability and observability Gramians, *respectively, that is, $P^c(P^a)$ and $Q^c(Q^a)$ are the unique positive semi-definite solutions to the Lyapunov equations*

$$AP^c + P^c A^T + BB^T = 0 \ \left(AP^a + P^a A^T - BB^T = 0\right) \tag{1.22}$$

$$A^T Q^c + Q^c A + C^T C = 0 \ \left(A^T Q^a + Q^a A - C^T C = 0\right). \tag{1.23}$$

Furthermore, if (A, B, C, D) is minimal then both Gramians are positive definite. □

1.4. BALANCED FORMS

From (1.22), (1.23) we rapidly deduce that if $(A, B, C, D) \sim \left(\widehat{A}, \widehat{B}, \widehat{C}, \widehat{D}\right)$ and S is the coordinate change $\hat{x} = Sx$ then $\widehat{P}^c = SP^cS^T$, $\widehat{Q}^c = S^{-T}Q^cS^{-1}$, and similarly for $\widehat{P}^a$ and $\widehat{Q}^a$.

Theorem 1.6 *Let T as in (1.17) be minimal and stable. Then there exists a coordinate change $\hat{x} = Sx$ such that*

$$\widehat{P}^c = S^T P^c S = \widehat{Q}^c = S^{-T} Q^c S^{-1} = \Sigma \tag{1.24}$$

where

$$\Sigma = \begin{bmatrix} \sigma_1 & & & \\ & \sigma_2 & & \\ & & \ddots & \\ & & & \sigma_n \end{bmatrix} \tag{1.25}$$

with $\overline{\sigma} := \sigma_1 \geq \sigma_2 \geq ... \geq \sigma_n =: \underline{\sigma} > 0.$ □

The equivalent $\left(\widehat{A}, \widehat{B}, \widehat{C}, \widehat{D}\right)$ of (A, B, C, D), via the transformation S introduced in the above theorem, is called the *balanced* form of the original system (A, B, C, D). Since $\widehat{P}^c\widehat{Q}^c = TP^cQ^cT^{-1} = \Sigma^2$, clearly $\sigma_1^2, ..., \sigma_n^2$ are the eigenvalues of P^cQ^c, and, in accordance with (2) of Theorem 1.5, $\sigma_1, ..., \sigma_n$ are called the *Hankel singular values* of T.

The construction of S is made according to the following algorithm:

1. Factorize Cholesky Q^c, *i.e.*,

$$Q^c = U^TU;$$

2. Bring UP^cU^T into the Schur form, that is,

$$UP^cU^T = V^T\Sigma^2V$$

where V is orthogonal and $\Sigma^2 = \text{diag}\left(\sigma_1^2, ..., \sigma_n^2\right)$;

3. Let

$$S := \Sigma^{-\frac{1}{2}}VU$$

which is the desired transformation.

Theorem 1.7 *Let T as in (1.17) be minimal and stable and let $\left(\widehat{A}, \widehat{B}, \widehat{C}, \widehat{D}\right)$ be the corresponding balanced form. Let also $\sigma_1 \geq ... \geq \sigma_n > 0$ be the Hankel singular values and assume the existence of an integer k for which $\sigma_{k+1} > \sigma_k$. Let Σ be partitioned as*

$$\Sigma = \begin{bmatrix} \Sigma_1 & \\ & \Sigma_2 \end{bmatrix} \tag{1.26}$$

where $\Sigma_1 = \text{diag}(\sigma_1, ..., \sigma_k)$, $\Sigma_2 = \text{diag}(\sigma_{k+1}, ..., \sigma_n)$ and let also $\widehat{A}, \widehat{B}, \widehat{C}$ be partitioned in accordance with Σ in (1.26), i.e.

$$\begin{aligned} \widehat{A} &= \begin{bmatrix} A_{11} & A_{12} \\ A_{21} & A_{22} \end{bmatrix}, \\ \widehat{B} &= \begin{bmatrix} B_1 \\ B_2 \end{bmatrix}, \\ \widehat{C} &= \begin{bmatrix} C_1 & C_2 \end{bmatrix}. \end{aligned} \tag{1.27}$$

Then both A_{11}, A_{22} *are stable and*

$$\begin{aligned} \|T - T_{11}\|_H^c &\leq \|T - T_{11}\|_\infty \\ &\leq 2 \sum_{i=k+1}^{n} \sigma_i , \end{aligned}$$

where

$$T_{11} = \left[\begin{array}{c|c} A_{11} & B_1 \\ \hline C_1 & D \end{array} \right].$$

□

T_{11} is called the *truncated* form of T with the tolerance $\gamma = 2\sum_{i=k+1}^{n} \sigma_i$.

NOTES AND REFERENCES

Basic definitions and general results on linear continuous-time systems can be found for instance in (Kailath, 1980). Concerning the Hankel and Toeplitz operators, more details are given in (Francis, 1986), (Glover, 1984) and (Partington, 1988). For results on exponential dichotomy in the continuous time case, see (Coppel, 1975). Leading ideas referring to causal (anticausal) operators may be found in (Bart *et al.*, 1979). Details concerning the balanced realizations with respect to the Gramians are given in (Glover, 1984).

CHAPTER 2

THE KALMAN–POPOV–YAKUBOVICH SYSTEM OF INDEFINITE SIGN

The present chapter stores the principal theoretical arguments on which the present monograph is built. The core of the subsequent development is the so called *generalized Popov–Yakubovich theory* — a natural extension of the famous positivity theory originated by Popov in the early 1960s. To be more specific, two leading facts will futher be pointed out. The first one concerns the existence conditions of the stabilizing solution to the Kalman–Popov–Yakubovich system (KPYS) in 'J-form' with weakest assumptions imposed on its coefficients. The second fact reveals a particular existence condition, called the *signature condition*, to which reduce the existence conditions of the solutions to all the problems formulated in this book.

2.1. POPOV TRIPLETS: ASSOCIATED OBJECTS, EQUIVALENCE, DUALITY

The standard framework for our study of the KPYS is the Popov triplet.

Definition 2.1 *A* Popov triplet *is a triplet of the form* $\Sigma = (A, B; P)$, *where* $A \in \mathbf{R}^{n\times n}$, $B \in \mathbf{R}^{n\times m}$, *and*

$$P = \begin{bmatrix} Q & L \\ L^T & R \end{bmatrix} = P^T \in \mathbf{R}^{(n+m)\times(n+m)}.$$

□

We shall also use the more elaborate notation $\Sigma = (A, B; Q, L, R)$. The Popov triplet should be regarded as a synthetic notation for a pair consisting of an initial value problem and a quadratic cost functional. In our case the pair is

$$\dot{x} = Ax + Bu, \qquad x(0) = \xi, \tag{2.1}$$

$$J_\Sigma(\xi, u) = \int_0^\infty \begin{bmatrix} x(t) \\ u(t) \end{bmatrix}^T P \begin{bmatrix} x(t) \\ u(t) \end{bmatrix} dt, \tag{2.2}$$

where (2.2) is defined for those $u \in \mathrm{L}_+^{2,m}$ for which (2.1) has a (unique) solution $x = x^{(\xi,u)} \in \mathrm{L}_+^{2,n}$.

The following mathematical objects associated with Σ will be now introduced.

(a) The Popov index

Let the set

$$U_\xi = \left\{u \in \mathrm{L}_+^{2,m}\text{: there exists a (unique) solution } x = x^{(\xi,u)} \in \mathrm{L}_+^{2,n} \text{ to (2.1)}\right\}$$

be introduced. Then the functional $J_\Sigma(\xi,.) : U_\xi \to \mathbf{R}$ defined through (2.2), where $x = x^{(\xi,u)}$, is called the *(usual) Popov index.* Notice that if A is *stable* then $U_\xi = \mathrm{L}_+^{2,m}$ for all ξ. If the pair (A, B) is *stabilizable*, then let $\widetilde{F}$ be such that $\widetilde{A} = A + B\widetilde{F}$ is stable and let $\tilde{x}^{(\xi,\tilde{u})}$ be the unique solution in $\mathrm{L}_+^{2,n}$ of the initial value problem $\dot{x} = \widetilde{A}x + B\tilde{u}$, $x(0) = \xi$. Then

$$U_\xi = \left\{u : u = \widetilde{F}\tilde{x}^{(\xi,\tilde{u})} + \tilde{u} \quad \forall\, \tilde{u} \in \mathrm{L}_+^{2,m}\right\} \quad \forall \xi \in \mathbf{R}^n$$

as easily can be checked.

Let A be *dichotomous.* Then for each $u \in \mathrm{L}^{2,m}$ the differential equation $\dot{x} = Ax + Bu$ has a unique solution $x_e^u \in \mathrm{L}^{2,n}$ and the functional

$$J_{\Sigma,e} : \mathrm{L}^{2,m} \to \mathbf{R}$$

introduced by

$$J_{\Sigma,e}(u) = \int_{-\infty}^{\infty} \begin{bmatrix} x_e^u(t) \\ u(t) \end{bmatrix}^T P \begin{bmatrix} x_e^u(t) \\ u(t) \end{bmatrix} dt \tag{2.3}$$

is well defined. We shall term $J_{\Sigma,e}(.)$ as the *extended* Popov index. Let $x^u = P_+^n x_e^u$. Then $J_\Sigma(.) : \mathrm{L}_+^{2,m} \to \mathbf{R}$ made explicit by

$$J_\Sigma(u) = \int_0^{\infty} \begin{bmatrix} x^u(t) \\ u(t) \end{bmatrix}^T P \begin{bmatrix} x^u(t) \\ u(t) \end{bmatrix} dt \qquad \left(u \in \mathrm{L}_+^{2,m}\right) \tag{2.4}$$

is the *reduced form* of the extended Popov index.

(b) The Popov function

Assume that A in Σ is dichotomous. Then (2.3) can be written in the equivalent form

$$J_{\Sigma,e}(u) = \left\langle \begin{bmatrix} x_e^u \\ u \end{bmatrix}, P \begin{bmatrix} x_e^u \\ u \end{bmatrix} \right\rangle_{\mathrm{L}^{2,n}\times\mathrm{L}^{2,m}} .$$

If $u \in \mathrm{L}^{2,m}$ denote by $\hat{u}$ its Fourier transform. Then, according to the Parseval identity, we obtain

$$J_{\Sigma,e}(u) = \frac{1}{2\pi} \left\langle \begin{bmatrix} \hat{x}_e^u \\ \hat{u} \end{bmatrix}, P \begin{bmatrix} \hat{x}_e^u \\ \hat{u} \end{bmatrix} \right\rangle_{\mathrm{L}^{2,n}\times\mathrm{L}^{2,m}} . \tag{2.5}$$

As $\hat{x}_e^u = (j\omega I - A)^{-1} B\hat{u}$, (2.5) gives

$$\begin{aligned} J_{\Sigma,e}(u) &= \frac{1}{2\pi}\left\langle \hat{u}, \begin{bmatrix} B^T\left(-j\omega I - A^T\right)^{-1} & I \end{bmatrix}\right. \\ &\qquad \left. \times \begin{bmatrix} Q & L \\ L^T & R \end{bmatrix} \begin{bmatrix} (j\omega I - A)^{-1} B \\ I \end{bmatrix} \hat{u} \right\rangle_{\mathrm{L}^{2,m}} . \end{aligned}$$

Let

$$\Pi_\Sigma(s) := \begin{bmatrix} B^T\left(-sI - A^T\right)^{-1} & I \end{bmatrix} \begin{bmatrix} Q & L \\ L^T & R \end{bmatrix} \begin{bmatrix} (sI - A)^{-1} B \\ I \end{bmatrix} \qquad (2.6)$$

be introduced for all $s \in \mathbf{C}\backslash\Lambda(A) \cup \Lambda(-A)$. Then one obtains

$$J_{\Sigma,e}(u) = \frac{1}{2\pi}\langle \hat{u}, \Pi_\Sigma(j\omega)\hat{u}\rangle_{\mathrm{L}^{2,m}} . \qquad (2.7)$$

Let A be arbitrary. Then the function Π_Σ, made explicit by (2.6) and defined everywhere in $\mathbf{C}$ except $\Lambda(A)\cup\Lambda(-A)$, is termed the *Popov function* associated with Σ.

It is a matter of elementary computations to prove the result just stated below.

Proposition 2.1 *The Popov function* Π_Σ *is realized via*

$$\Pi_\Sigma := \left[\begin{array}{cc|c} A & 0 & B \\ -Q & -A^T & -L \\ \hline L^T & B^T & R \end{array}\right]. \qquad (2.8)$$

□

As we shall see, the Popov function will play a key role throughout this book.

(c) The Kalman–Popov–Yakubovich system in J-form (KPYS(Σ, J))

The following nonlinear system in the unknowns $X \in \mathbf{R}^{n\times n}$, $V \in \mathbf{R}^{m\times m}$ and $W \in \mathbf{R}^{m\times n}$ with X *symmetric*

$$\begin{aligned} R &= V^T J V \\ L + XB &= W^T J V \\ Q + A^T X + XA &= W^T J W \end{aligned} \qquad (2.9)$$

where J is any sign matrix, *i.e.*, it is of form $J = \begin{bmatrix} -I_{m_1} & 0 \\ 0 & I_{m_2} \end{bmatrix}$ ($m = m_1 + m_2$), is called the Kalman–Popov–Yakubovich system in J-form and will be abbreviated KPYS(Σ, J). If R is nonsingular then V is nonsingular too and $\operatorname{sgn} R = J$ automatically.

Assume now that R in Σ is nonsingular. Then any triple (X, V, W) satisfying (2.9) is called a *(anti-) stabilizing solution* if

$$F = -V^{-1}W \tag{2.10}$$

makes $A + BF$ (anti-) stable. In this case (2.9) reduces via (2.10) to

$$\begin{bmatrix} A^TX + XA + Q & L + XB \\ L^T + B^TX & R \end{bmatrix} \begin{bmatrix} I \\ F \end{bmatrix} = 0 \tag{2.11}$$

as can be directly checked by combining (2.9) with (2.10). The uniqueness of the stabilizing solution to the KPYS(Σ, J) is emphasized by the next proposition:

Proposition 2.2 *Assume that R is nonsingular in (2.9). If (X, V, W) and $\left(\widetilde{X}, \widetilde{V}, \widetilde{W}\right)$ are two (anti-) stabilizing solution to (2.9) then $X = \widetilde{X}$.*

Proof. From (2.11) one deduces

$$\begin{aligned} A^TX + X(A + BF) + LF + Q &= 0 \\ L^T + B^TX + RF &= 0. \end{aligned}$$

Let $\widetilde{F} = -\widetilde{V}^{-1}\widetilde{W}$. Then by adding to the first equation the second premultiplied by $\widetilde{F}^T$ one obtains

$$\left(A + B\widetilde{F}\right)^T X + X(A + BF) + LF + \widetilde{F}^TL^T + \widetilde{F}^TRF = 0. \tag{2.12}$$

Similarly, if (2.11) is updated with $\left(\widetilde{X}, \widetilde{V}, \widetilde{W}\right)$ one obtains

$$(A + BF)^T \widetilde{X} + \widetilde{X}\left(A + B\widetilde{F}\right) + L\widetilde{F} + F^TL^T + F^TR\widetilde{F} = 0$$

and by transposition

$$\left(A + B\widetilde{F}\right)^T \widetilde{X} + \widetilde{X}(A + BF) + \widetilde{F}^TL^T + LF + \widetilde{F}^TRF = 0. \tag{2.13}$$

Subtracting (2.13) from (2.12) one obtains eventually

$$\left(A + B\widetilde{F}\right)^T \left(X - \widetilde{X}\right) + \left(X - \widetilde{X}\right)(A + BF) = 0.$$

As both $A + BF$ and $A + B\widetilde{F}$ are stable, the above Lyapunov equation reveals that $X - \widetilde{X} = 0$, *i.e.* $X = \widetilde{X}$ and the conclusion follows. For the paranthesised text the proof runs similarly. □

Remark 2.1 Let R be nonsingular and assume the existence of a stabilizing solution (X, V, X) to the KPYS(Σ, J) (2.9). As V is nonsingular, both V and W can be eliminated in (2.9) and we obtain that X satisfies the *algebraic Riccati equation* ARE(Σ)

$$A^T X + XA - (L + XB) R^{-1} \left(B^T X + L^T\right) + Q = 0. \tag{2.14}$$

Further, F given by (2.10) becomes

$$F = -R^{-1} \left(B^T X + L^T\right). \tag{2.15}$$

Since $A + BF$ is stable X is called the *stabilizing solution* of the ARE(Σ) (2.14). Conversely, if (2.14) is fulfilled for some symmetric X with the additional property that $A + BF$ is stable for F given by (2.15) then, for $J = \operatorname{sgn} R$, the KPYS(Σ, J) has a stabilizing solution (X, V, W). Indeed, if we J-factorize R as $R = V^T JV$ and then we define $W^T = (L + XB)V^{-1}J$, the conclusion follows trivially. Furthermore, in the light of Proposition 2.1 one concludes that if the ARE(Σ) has a stabilizing solution then it is unique. □

Proposition 2.3 *Assume that (X, V, W) is a solution of the KPYS(Σ, J) (2.9), not necessarily a stabilizing one. Then*

$$J_\Sigma(\xi, u) = \xi^T X \xi + \left\langle Vu + Wx^{(\xi,u)},\, J\left(Vu + Wx^{(\xi,u)}\right)\right\rangle_{\mathrm{L}_+^{2,m}} \tag{2.16}$$

for all $u \in U_\xi$. If in addition R is nonsingular and (X, V, W) is a stabilizing solution then $J_\Sigma(\xi)$ has a stationary point in U_ξ expressed in the feedback form $u = Fx$, where F is given via (2.10). If futhermore A is dichotomous then

$$J_{\Sigma,e}(u) = \langle Vu + Wx_e^u,\, J\,(Vu + Wx_e^u)\rangle_{\mathrm{L}^{2,m}} \quad \forall\, u \in \mathrm{L}^{2,m} \tag{2.17}$$

and, in particular,

$$\begin{aligned} J_\Sigma(u) &= \langle Vu + Wx^u,\, J\,(Vu + Wx^u)\rangle_{\mathrm{L}_+^{2,m}} \\ &\quad + x^T(0)Xx(0) \qquad \forall\, u \in \mathrm{L}_+^{2,m}. \end{aligned} \tag{2.18}$$

Proof. Using (2.9) we can write for each $u \in U_\xi$ and corresponding $x = x^{(\xi,u)}$

$$\begin{bmatrix} x \\ u \end{bmatrix}^T \begin{bmatrix} Q & L \\ L^T & R \end{bmatrix} \begin{bmatrix} x \\ u \end{bmatrix}$$

$$
\begin{aligned}
&= \begin{bmatrix} x \\ u \end{bmatrix}^T \begin{bmatrix} W^TJW & W^TJV \\ V^TJW & V^TJV \end{bmatrix} \begin{bmatrix} x \\ u \end{bmatrix} \\
&\quad - \begin{bmatrix} x \\ u \end{bmatrix}^T \begin{bmatrix} A^TX+XA & XB \\ B^TX & 0 \end{bmatrix} \begin{bmatrix} x \\ u \end{bmatrix} \\
&= (Wx+Vu)^T J (Wx+Vu) - (Ax+Bu)^T Xx \\
&\quad -x^T X (Ax+Bu) \\
&= (Wx+Vu)^T J (Wx+Vu) - \dot{x}^T Xx - x^T X\dot{x} \\
&= (Wx+Vu)^T J (Wx+Vu) - \frac{d}{dt}\left(x^T Xx\right). \qquad (2.19)
\end{aligned}
$$

By integrating both leftmost and rightmost terms in (2.19) from 0 to ∞ one obtains (2.16). Furthermore, if in addition R is nonsingular and (X, V, W) is a stabilizing solution one can easily notice that the second term in (2.16) vanishes for $u = Fx$. Applying the same machinery the rest of the proof follows trivially. □

(d) The extended Hamiltonian pencil (EHP(Σ))

The matrix pencil $\lambda M - N$ defined by

$$
M = \begin{bmatrix} I_n & 0 & 0 \\ 0 & I_n & 0 \\ 0 & 0 & 0 \end{bmatrix}, \quad N = \begin{bmatrix} A & 0 & B \\ -Q & -A^T & -L \\ L^T & B^T & R \end{bmatrix}, \qquad (2.20)
$$

with $M,\ N \in \mathbf{R}^{(2n+m)\times(2n+m)}$ is called the *extended Hamiltonian pencil* associated with Σ.

By comparing (2.20) with (2.8) one concludes that the transmission zeros (finite and infinite) of the realization (2.8) coincide with the spectrum (finite and infinite) of the EHP (2.20).

Definition 2.2 *Let* $\Sigma = (A, B; Q, L, R)$ *and* $\widetilde{\Sigma} = \left(\widetilde{A}, \widetilde{B}; \widetilde{Q}, \widetilde{L}, \widetilde{R}\right)$ *be two Popov triplets. We shall say that* Σ *is* equivalent *to* $\widetilde{\Sigma}$ *and we shall write* $\Sigma \sim \widetilde{\Sigma}$, *if there exist* $\widetilde{X} = \widetilde{X}^T \in \mathbf{R}^{n\times n}$ *and* $\widetilde{F} \in \mathbf{R}^{m\times n}$ *such that*

$$
\begin{aligned}
\widetilde{A} &= A + B\widetilde{F}, \\
\widetilde{B} &= B, \\
\widetilde{Q} &= Q + L\widetilde{F} + \widetilde{F}^T L^T + \widetilde{F}^T R\widetilde{F} + \widetilde{A}^T\widetilde{X} + \widetilde{X}\widetilde{A}, \qquad (2.21) \\
\widetilde{L} &= L + \widetilde{F}^T R + \widetilde{X}B, \\
\widetilde{R} &= R.
\end{aligned}
$$

If $\widetilde{X} = 0$ *then* $\widetilde{\Sigma}$ *is termed an* $\widetilde{F}$-equivalent *of* Σ. *If* $\widetilde{F} = 0$ *and* $\widetilde{Q} = 0$ *then* $\widetilde{\Sigma}$ *is called a* reduced equivalent *to* Σ. □

One can prove easily that $\sim$ is actually an equivalence relation.

Remark 2.2 If the pair (A, B) is stabilizable (antistabilizable) then there exists a unique $\widetilde{X}$ for which $\widetilde{Q} = 0$. The same assertion holds either for (A, B) controllable or A stable (antistable). Indeed, if $\widetilde{F}$ is such that $\widetilde{A} = A + B\widetilde{F}$ is stable (antistable) then clearly there exists a unique symmetric $\widetilde{X}$ for which the right hand side of the third equation in (2.21) vanishes. In fact $\widetilde{X}$ is the unique solution of an appropriate Lyapunov equation. □

If $\widetilde{\Sigma}$ is any equivalent to Σ then the objects associated with $\widetilde{\Sigma}$ are related to those associated with Σ through the proposition stated below

Proposition 2.4 *Let Σ be given and let $\widetilde{\Sigma}$ be any equivalent of Σ. Then the following hold*

(1)

(a)

$$J_{\Sigma}(\xi, u) = \xi^T \widetilde{X} \xi + J_{\widetilde{\Sigma}}(\xi, \tilde{u})$$

for $\tilde{u} = u - \widetilde{F} x^{(\xi,u)}$ *and* $\forall\, u \in U_{\xi}$;

(b)

$$J_{\Sigma,e}(u) = J_{\widetilde{\Sigma},e}(\tilde{u})$$

for $\tilde{u} = u - F x_e^u$ *and* $\forall\, u \in \mathrm{L}^{2,m}$;

(c)

$$J_{\Sigma}(u) = J_{\widetilde{\Sigma}}(\tilde{u}) + x^T(0)\widetilde{X}x(0)$$

for $\tilde{u} = u - F x^u$ *and* $\forall\, u \in \mathrm{L}_+^{2,m}$;

(2)

$$\Pi_{\Sigma}(s) = S_{\widetilde{F}}^*(s)\Pi_{\widetilde{\Sigma}}(s)S_{\widetilde{F}}(s)$$

where

$$S_{\widetilde{F}}(s) := \left[\begin{array}{c|c} A & B \\ \hline -\widetilde{F} & I \end{array}\right], \quad S_{\widetilde{F}}^{-1}(s) := \left[\begin{array}{c|c} \widetilde{A} & B \\ \hline \widetilde{F} & I \end{array}\right].$$

If in addition (2.11) holds then

$$\Pi_{\Sigma}(s) = S_{\widetilde{F}}^*(s) R S_{\widetilde{F}}(s) \tag{2.22}$$

and (2.22) is usually known as the spectral factorization identity.

(3) (X, V, W) is a solution to the KPYS(Σ, J) if and only if

$$\left(X - \widetilde{X}, V, W + V\widetilde{F}\right)$$

is a solution to the KPYS($\widetilde{\Sigma}, J$). In particular, if R is nonsingular then X is a symmetric (stabilizing) solution to the ARE(Σ) if and only if $X - \widetilde{X}$ is a symmetric (stabilizing) solution to the ARE($\widetilde{\Sigma}$).

(4) The EHP(Σ) $\lambda M - N$ and EHP($\widetilde{\Sigma}$) $\lambda\widetilde{M} - \widetilde{N}$ are strictly equivalent, i.e., there exist $\widetilde{U}, \widetilde{W} \in \mathbf{R}^{(2n+m)\times(2n+m)}$, both nonsingular, such that $\lambda M - N = \widetilde{U}(\lambda\widetilde{M} - \widetilde{N})\widetilde{W}$.

Proof. (1a) For each ξ and $u \in U_\xi$ with $U_\xi \neq \phi$ we can write

$$\begin{aligned}
&\begin{bmatrix} x^{(\xi,u)} \\ u \end{bmatrix}^T \begin{bmatrix} Q & L \\ L^T & R \end{bmatrix} \begin{bmatrix} x^{(\xi,u)} \\ u \end{bmatrix} \\
&= \begin{bmatrix} x^{(\xi,u)} \\ u \end{bmatrix}^T \begin{bmatrix} I & -\widetilde{F}^T \\ 0 & I \end{bmatrix} \begin{bmatrix} I & \widetilde{F}^T \\ 0 & I \end{bmatrix} \begin{bmatrix} Q & L \\ L^T & R \end{bmatrix} \\
&\quad \times \begin{bmatrix} I & 0 \\ \widetilde{F} & I \end{bmatrix} \begin{bmatrix} I & 0 \\ -\widetilde{F} & I \end{bmatrix} \begin{bmatrix} x^{(\xi,u)} \\ u \end{bmatrix} \\
&= \begin{bmatrix} x^{(\xi,u)} \\ u \end{bmatrix}^T \begin{bmatrix} \widetilde{Q} - \widetilde{A}^T\widetilde{X} - \widetilde{X}\widetilde{A} & \widetilde{L} - \widetilde{X}B \\ \widetilde{L}^T - B^T\widetilde{X} & R \end{bmatrix} \begin{bmatrix} x^{(\xi,u)} \\ u \end{bmatrix} \\
&= \begin{bmatrix} \tilde{x}^{(\xi,\tilde{u})} \\ \tilde{u} \end{bmatrix}^T \begin{bmatrix} \widetilde{Q} & \widetilde{L} \\ \widetilde{L}^T & R \end{bmatrix} \begin{bmatrix} \tilde{x}^{(\xi,\tilde{u})} \\ \tilde{u} \end{bmatrix} - \frac{d}{dt}\left(\tilde{x}^{(\xi,\tilde{u})T}\widetilde{X}\tilde{x}^{(\xi,\tilde{u})}\right), \quad (2.23)
\end{aligned}$$

where $\tilde{x}^{(\xi,\tilde{u})} \equiv x^{(\xi,u)}$ because of $\dot{x} = Ax + Bu = \widetilde{A}x + B\tilde{u}$, $x(0) = \xi$, $u \in U_\xi$.

By integrating the leftmost and the rightmost terms in (2.23) from 0 to ∞, (1a) follows. To prove (1b) and (1c) we proceed similarly.

(2)

$$\begin{aligned}
\Pi_\Sigma(s) &= \begin{bmatrix} B^T\left(-sI - A^T\right)^{-1} & I \end{bmatrix} \begin{bmatrix} Q & L \\ L^T & R \end{bmatrix} \begin{bmatrix} (sI - A)^{-1}B \\ I \end{bmatrix} \\
&= \begin{bmatrix} B^T\left(-sI - A^T\right)^{-1} & I \end{bmatrix} \begin{bmatrix} I & -\widetilde{F}^T \\ 0 & I \end{bmatrix} \begin{bmatrix} I & \widetilde{F}^T \\ 0 & I \end{bmatrix} \\
&\quad \times \begin{bmatrix} Q & L \\ L^T & R \end{bmatrix} \begin{bmatrix} I & 0 \\ \widetilde{F} & I \end{bmatrix} \begin{bmatrix} I & 0 \\ -\widetilde{F} & I \end{bmatrix} \begin{bmatrix} (sI - A)^{-1}B \\ I \end{bmatrix} \\
&= \begin{bmatrix} B^T\left(-sI - A^T\right)^{-1} & S^*_{\widetilde{F}}(s) \end{bmatrix}^T \\
&\quad \times \begin{bmatrix} \widetilde{Q} - \widetilde{A}^T\widetilde{X} - \widetilde{X}\widetilde{A} & \widetilde{L} - \widetilde{X}B \\ \widetilde{L}^T - B^T\widetilde{X} & R \end{bmatrix} \begin{bmatrix} (sI - A)^{-1}B \\ S_{\widetilde{F}}(s) \end{bmatrix} \\
&= S^*_{\widetilde{F}}(s) \begin{bmatrix} B^T\left(-sI - \widetilde{A}^T\right)^{-1} & I \end{bmatrix} \begin{bmatrix} \widetilde{Q} & \widetilde{L} \\ \widetilde{L}^T & R \end{bmatrix}
\end{aligned}$$

$$\times \begin{bmatrix} \left(sI-\widetilde{A}\right)^{-1} B \\ I \end{bmatrix} S_{\widetilde{F}}(s) - S^*_{\widetilde{F}}(s) \begin{bmatrix} B^T \left(-sI-\widetilde{A}\right)^{-1} & I \end{bmatrix}$$

$$\times \begin{bmatrix} \widetilde{A}^T\widetilde{X}+\widetilde{X}\widetilde{A} & \widetilde{X}B \\ B^T\widetilde{X} & 0 \end{bmatrix} \begin{bmatrix} \left(sI-\widetilde{A}\right)^{-1} B \\ I \end{bmatrix} S_{\widetilde{F}}(s). \qquad (2.24)$$

But

$$\begin{aligned}
&\begin{bmatrix} B^T \left(-sI-\widetilde{A}^T\right)^{-1} & I \end{bmatrix} \begin{bmatrix} \widetilde{A}^T\widetilde{X}+\widetilde{X}\widetilde{A} & \widetilde{X}B \\ B^T\widetilde{X} & 0 \end{bmatrix} \begin{bmatrix} \left(sI-\widetilde{A}\right)^{-1} B \\ I \end{bmatrix} \\
&= B^T\left(-sI-\widetilde{A}^T\right)^{-1}\left(\widetilde{A}^T\widetilde{X}+\widetilde{X}\widetilde{A}\right)\left(sI-\widetilde{A}\right)^{-1}B \\
&\quad +B^T\left(-sI-\widetilde{A}^T\right)^{-1}\widetilde{X}B+B^T\widetilde{X}\left(sI-\widetilde{A}\right)^{-1}B \\
&= B^T\left(-sI-\widetilde{A}^T\right)^{-1}\left(\widetilde{A}^T\widetilde{X}\left(sI-\widetilde{A}\right)^{-1}+\widetilde{X}\right)B \\
&\quad +B^T\left(\left(-sI-\widetilde{A}\right)^{-1}\widetilde{X}\widetilde{A}+\widetilde{X}\right)\left(sI-\widetilde{A}\right)^{-1}B \\
&= B^T\left(-sI-\widetilde{A}^T\right)^{-1}\left(\widetilde{A}^T\widetilde{X}-\widetilde{X}A+s\widetilde{X}\right)\left(sI-\widetilde{A}\right)^{-1}B \\
&\quad +B^T\left(-sI-\widetilde{A}^T\right)^{-1}\left(\widetilde{X}\widetilde{A}-\widetilde{A}^T\widetilde{X}-s\widetilde{X}\right)\left(sI-\widetilde{A}\right)^{-1}B \\
&= 0.
\end{aligned}$$

Using this last identity in (2.24) the conclusion follows.
To prove (2.22) use (2.14) and (2.15) in (2.21) and obtain

$$\begin{aligned}
\widetilde{Q} &= Q-LR^{-1}\left(B^TX+L^T\right)-(L+XB)R^{-1}L^T \\
&\quad +(L+XB)R^{-1}\left(B^TX+L^T\right)+A^TX+XA \\
&\quad -(L+XB)R^{-1}B^TX-XBR^{-1}\left(L^T+B^TX\right) \\
&= 0
\end{aligned}$$

and

$$\widetilde{L}=L-(L+XB)+XB=0.$$

Therefore $\Pi_{\widetilde{\Sigma}} = R$ in accordance with the updated version of (2.6), and (2.22) follows from the identity that we have just proved.

(3) Using (2.9) combined with (2.21) we shall write

$$\begin{aligned}
\widetilde{L}+\left(X-\widetilde{X}\right)B &= L+\widetilde{F}^TR+XB \\
&= W^TJV+\widetilde{F}^TR
\end{aligned}$$

$$\begin{aligned} &= W^T J V + \widetilde{F}^T V^T J V \\ &= \left(W + V\widetilde{F}\right)^T J V \end{aligned}$$

and

$$\begin{aligned} &\widetilde{Q} + \widetilde{A}^T\left(X - \widetilde{X}\right) + \left(X - \widetilde{X}\right)\widetilde{A} \\ &= Q + L\widetilde{F} + \widetilde{F}^T L^T + \widetilde{F}^T R\widetilde{F} + A^T X + XA + \widetilde{F}^T B^T X + XB\widetilde{F} \\ &= W^T J W + (L + XB)\widetilde{F} + \widetilde{F}^T\left(L^T + B^T X\right) + \widetilde{F}^T R\widetilde{F} \\ &= W^T J W + \left(L + XB + \widetilde{F}^T R\right)\widetilde{F} + \widetilde{F}^T\left(L^T + B^T X + R\widetilde{F}\right) - \widetilde{F}^T R\widetilde{F} \\ &= W^T J W + \left(W + V\widetilde{F}\right)^T J V\widetilde{F} + \widetilde{F}^T V^T J\left(W + V\widetilde{F}\right) - \widetilde{F}^T V^T J V\widetilde{F} \\ &= \left(W + V\widetilde{F}\right)^T J\left(W + V\widetilde{F}\right) \end{aligned}$$

and the assertion is proved.

(4) If we take

$$\widetilde{U} = \begin{bmatrix} I & 0 & 0 \\ \widetilde{X} & I & \widetilde{F}^T \\ 0 & 0 & I \end{bmatrix}, \quad \widetilde{W} = \begin{bmatrix} I & 0 & 0 \\ -\widetilde{X} & I & 0 \\ -\widetilde{F} & 0 & I \end{bmatrix}$$

then simple computations prove the strict equivalence between the EHP(Σ) and EHP($\widetilde{\Sigma}$). □

Definition 2.3 *Let* $\Sigma = (A, B; Q, L, R)$ *be a Popov triplet and assume that it has a reduced equivalent* $\widetilde{\Sigma} = \left(A, B; 0, \widetilde{L}, R\right)$. *Then the Popov triplet* $\Sigma_d = \left(-A^T, -\widetilde{L}; 0, B, R\right)$ *is termed as the* dual *of* Σ. □

Remark 2.3 According to Remark 2.2 if A in Σ is stable (antistable) then Σ has always a dual in which its first entry is antistable (stable). □

Proposition 2.5 *If* Σ *has a dual* Σ_d *then*

$$\Pi_\Sigma = \Pi_{\Sigma_d}. \tag{2.25}$$

Proof. Let $\widetilde{\Sigma}$ be the reduced equivalent of Σ. Then in the light of (2) of Proposition 2.4 one obtains

$$\begin{aligned} &\Pi_\Sigma(s) = \Pi_{\widetilde{\Sigma}}(s) \\ &= R + \widetilde{L}^T(sI - A)^{-1}B + B^T\left(-sI - A^T\right)^{-1}\widetilde{L} \\ &= R + B^T\left(sI - \left(-A^T\right)\right)^{-1}\left(-\widetilde{L}\right) + \left(-\widetilde{L}\right)^T\left(-sI - \left(-A^T\right)^T\right)^{-1} \\ &= \Pi_{\Sigma_d}(s). \end{aligned}$$

□

2.2. BASIC OPERATORS

According to Theorem 1.2, if A is dichotomous then the differential equation

$$\dot{x} = Ax + Bu \tag{2.26}$$

has for each $u \in L^{2,m}$ a unique solution $x_e^u \in L^{2,n}$ given explicitly by

$$x_e^u = \int_{-\infty}^{t} \Pi_- e^{A(t-\tau)} Bu(\tau) d\tau - \int_{t}^{\infty} \Pi_+ e^{A(t-\tau)} Bu(\tau) d\tau, \tag{2.27}$$

and there exists $\rho > 0$ such that

$$\|x_e^u\|_2 \leq \rho \|u\|_2 \qquad \forall\, u \in L^{2,m}. \tag{2.28}$$

Here Π_- (Π_+) is the projection on the stable (antistable) subspace χ_- (χ_+) associated with A, *i.e.*, $\mathbf{R}^n = \chi_- \oplus \chi_+$ and $\chi_\pm = \ker \mu_\pm(A)$ where $\mu(\lambda)$ is the minimal polynomial of A factorized as $\mu = \mu_+\mu_-$ with respect to the imaginary axis.

With other words (2.26) defines, under the dichotomy hypothesis, a linear bounded *input state* operator from $L^{2,m}$ to $L^{2,n}$ denoted $\mathcal{L}_e$. Thus the condensed forms of (2.27) and (2.28) are $x_e^u = \mathcal{L}_e u$ and $\|\mathcal{L}_e\| \leq \rho$, respectively.

Furthermore, if A is dichotomous then the extended Popov index $J_{\Sigma,e}$ can be written as

$$\begin{aligned} J_{\Sigma,e}(u) &= \left\langle u, \begin{bmatrix} \mathcal{L}_e^* & I \end{bmatrix} \begin{bmatrix} Q & L \\ L^T & R \end{bmatrix} \begin{bmatrix} \mathcal{L}_e \\ I \end{bmatrix} u \right\rangle_{L^{2,m}} \\ &= \langle u, \mathcal{R}_e u \rangle_{L^{2,m}}\,, \end{aligned} \tag{2.29}$$

where

$$\mathcal{R}_e := R + L^T \mathcal{L}_e + \mathcal{L}_e^* L + \mathcal{L}_e^* Q \mathcal{L}_e \tag{2.30}$$

is a linear bounded operator from $L^{2,m}$ into itself.

If we compare (2.7) with (2.29) we conclude that Π_Σ *is the symbol* of $\mathcal{R}_e$. With other words for each $u \in L^{2,m}$ we have $v = \mathcal{R}_e u \in L^{2,m}$ if and only if $\hat{v}(j\omega) = \Pi_\Sigma(j\omega)\hat{u}(j\omega)$ where $\hat{u}$ is the Fourier transform of u and $\hat{u} \in L^{2,m}$.

(A) Assume that A is stable

In this case if we consider the initial value problem (2.1) then for each $\xi \in \mathbf{R}^n$ and $u \in L_+^{2,m}$

$$x^{(\xi,u)} = \Phi\xi + \mathcal{L}u \tag{2.31}$$

is the unique L^2-solution. Here

$$(\Phi\xi)(t) = e^{At}\xi\,, \quad t \geq 0 \tag{2.32}$$

and

$$\mathcal{L} = P_+^n \mathcal{L}_e P_+^m = \mathcal{L}_e P_+^m \tag{2.33}$$

is the Toeplitz operator associated with $\mathcal{L}_e$. Notice that the last equality in (2.33) follows trivially from (2.27) because of $\Pi_- = I$ and $\Pi_+ = 0$, that is,

$$(\mathcal{L}u)(t) = (P_+^n x_e^u)(t) = x^u(t) = \int_0^t e^{A(t-\tau)} Bu(\tau) d\tau.$$

With (2.31) substituted in the Popov index (2.2) one obtains

$$\begin{aligned} J_\Sigma(\xi, u) &= \left\langle \begin{bmatrix} \xi \\ u \end{bmatrix}, \begin{bmatrix} \Phi^* & 0 \\ \mathcal{L}^* & I \end{bmatrix} \begin{bmatrix} Q & L \\ L^T & R \end{bmatrix} \begin{bmatrix} \Phi & \mathcal{L} \\ 0 & I \end{bmatrix} \begin{bmatrix} \xi \\ u \end{bmatrix} \right\rangle_{\mathbf{R}^n \times \mathrm{L}_+^{2,m}} \\ &= \xi^T \widetilde{X} \xi + 2 \langle u, \mathcal{P}^* \xi \rangle_{\mathrm{L}_+^{2,m}} + \langle u, \mathcal{R} u \rangle_{\mathrm{L}_+^{2,m}}, \end{aligned} \tag{2.34}$$

where

$$\widetilde{X} := \Phi^* Q \Phi, \tag{2.35}$$

$$\mathcal{P} := \Phi^* (Q\mathcal{L} + L), \tag{2.36}$$

$$\mathcal{R} := R + L^T \mathcal{L} + \mathcal{L}^* L + \mathcal{L}^* Q \mathcal{L}, \tag{2.37}$$

and the adjoints Φ^* and $\mathcal{L}^*$ of Φ and $\mathcal{L}$ are given by

$$\Phi^* x = \int_0^\infty e^{A^T t} x(t) dt \quad \forall x \in \mathrm{L}_+^{2,n}, \tag{2.38}$$

and

$$(\mathcal{L}^* x)(t) = \int_t^\infty B^T e^{A^T(\tau - t)} x(\tau) d\tau \quad \forall\, x \in \mathrm{L}_+^{2,n}, \tag{2.39}$$

respectively, as simple computations show.

One can rapidly recognize that $\widetilde{X}$ in (2.35) can be expressed via (2.38) as

$$\widetilde{X} = \int_0^\infty e^{A^T t} Q e^{At} dt, \tag{2.40}$$

that is, $\widetilde{X}$ is the unique symmetric solution to the Lyapunov equation

$$A^T \widetilde{X} + \widetilde{X} A + Q = 0. \tag{2.41}$$

Proposition 2.6 *Let $\Sigma = (A, B; Q, L, R)$ be a Popov triplet with A stable and let $\widetilde{\Sigma} = \left(A, B; 0, \widetilde{L}, R\right)$ be its reduced equivalent. If $\widetilde{\mathcal{R}}_e$, $\widetilde{\mathcal{P}}$ and $\widetilde{\mathcal{R}}$ are the operators (2.30), (2.36) and (2.37) associated with $\widetilde{\Sigma}$ then the following hold:*

(1)

$$\widetilde{\mathcal{R}}_e = \mathcal{R}_e;$$

(2)

$$\mathcal{R} = P_+^m \mathcal{R}_e P_+^m = P_+^m \widetilde{\mathcal{R}}_e P_+^m = \widetilde{\mathcal{R}},$$

i.e., $\mathcal{R}\left(\widetilde{\mathcal{R}}\right)$ *is the Toeplitz operator associated with* $\mathcal{R}_e\left(\widetilde{\mathcal{R}}_e\right)$*;*

(3)

$$\widetilde{\mathcal{P}} = \mathcal{P}.$$

Proof. Notice first that $\widetilde{L} = L + \widetilde{X}B$, where $\widetilde{X}$ is that for which (2.41) is fulfilled and has the explicit form given by (2.40).

(1) Since $\Pi_\Sigma = \Pi_{\widetilde{\Sigma}}$ (see (2) of Proposition 2.4 for $\widetilde{F} = 0$) the equality $\widetilde{\mathcal{R}}_e = \mathcal{R}_e$ follows trivially.

(2) Using the last equality in (2.33) which in this case is essential, the equality $\mathcal{R} = P_+^m \mathcal{R}_e P_+^m$ follows automatically. The next equalities are trivial.

(3) We can write

$$\xi^T \mathcal{P} u = \xi^T \Phi^* (Q\mathcal{L} + L) u = \xi^T \widetilde{\mathcal{P}} u + \xi^T \left(\Phi^* Q\mathcal{L} - \Phi^* \widetilde{X} B\right) u. \tag{2.42}$$

Let $x_f = \mathcal{L}u$ and $x_l = \Phi\xi$ be the forced and free solutions to the initial value problem (2.1), respectively. Then

$$\begin{aligned}
\xi^T \Phi^* Q\mathcal{L}u &= \langle \Phi\xi, Q\mathcal{L}u\rangle_{\mathrm{L}_+^{2,n}} = \langle x_l, Qx_f\rangle_{\mathrm{L}_+^{2,n}} \\
&= \left\langle x_l, \left(-A^T\widetilde{X} - \widetilde{X}A\right) x_f\right\rangle_{\mathrm{L}_+^{2,n}} \\
&= -\left\langle Ax_l, \widetilde{X}x_f\right\rangle_{\mathrm{L}_+^{2,n}} - \left\langle x_l, \widetilde{X}\left(\dot{x}_f - Bu\right)\right\rangle_{\mathrm{L}_+^{2,n}} \\
&= -\left\langle \dot{x}_l, \widetilde{X}x_f\right\rangle_{\mathrm{L}_+^{2,n}} - \left\langle x_l, \widetilde{X}\dot{x}_f\right\rangle_{\mathrm{L}_+^{2,n}} + \left\langle x_l, \widetilde{X}Bu\right\rangle_{\mathrm{L}_+^{2,n}} \\
&= -\int_0^\infty \frac{d}{dt}\left(x_l^T \widetilde{X} x_f\right) dt + \xi^T \Phi^* \widetilde{X} B u \\
&= \xi^T \Phi^* \widetilde{X} B u.
\end{aligned}$$

Hence $\xi^T \left(\Phi^* Q\mathcal{L} - \Phi^* \widetilde{X} B\right) u = 0$, from which (2.42) yields $\xi^T \mathcal{P} u = \xi^T \widetilde{\mathcal{P}} u$ $\forall \xi \in \mathbf{R}^n$ and $\forall u \in \mathrm{L}_+^{2,m}$. Thus $\widetilde{\mathcal{P}} = \mathcal{P}$. □

(B) Assume that A is antistable

In this case the reduced form of the extended Popov index $J_\Sigma(u)$ (2.4) is given by

$$J_\Sigma(u) = \langle u, \mathcal{R}u\rangle_{\mathrm{L}_+^{2,m}}, \tag{2.43}$$

where $\mathcal{R}$ has formally the same expression (2.37) but, in accordance with (2.27), in this case we have

$$(\mathcal{L}u)(t) = -\int_t^\infty e^{A(t-\tau)}Bu(\tau)d\tau, \quad t \geq 0, \tag{2.44}$$

and

$$\mathcal{L} = P_+^n \mathcal{L}_e P_+^m \tag{2.45}$$

is the Toeplitz operator associated with $\mathcal{L}_e$. Unlike the previous case the last equality in (2.33) *does not hold.* Thus

$$\mathcal{L} \neq \mathcal{L}_e P_+^m, \tag{2.46}$$

as one can deduce from (2.27). The last result is the so called *anticausal* counterpart of Proposition 2.6.

Proposition 2.7 *Let $\Sigma = (A, B; Q, L, R)$ be a Popov triplet with A antistable and let $\widetilde{\Sigma} = \left(A, B; 0, \widetilde{L}, R\right)$ be its reduced equivalent. Associated with Σ, let $\mathcal{R}$ be formally defined through (2.37), where $\mathcal{L}$ has been updated by (2.44), and let $\widetilde{\mathcal{R}}$ be associated with $\widetilde{\Sigma}$ with similar significance. Then the following hold:*

(1)

$$\widetilde{\mathcal{R}} = P_+^m \widetilde{\mathcal{R}}_e P_+^m = P_+^m \mathcal{R}_e P_+^m,$$

with $\mathcal{R}_e$ defined by (2.30).

(2)

$$\mathcal{R} = \widetilde{\mathcal{R}} + \widehat{\Psi}^* \widetilde{X} \widehat{\Psi},$$

where $\widetilde{X}$ is given by (2.41) in which A is now antistable and $\widehat{\Psi}$ is the anticausal controllability operator defined by

$$\widehat{\Psi}u = -\int_0^\infty e^{-At}Bu(t)dt. \tag{2.47}$$

Proof. (1) Since $\widetilde{\mathcal{R}}_e = R + \widetilde{L}^T \mathcal{L}_e + \mathcal{L}_e^* \widetilde{L}$ it follows that

$$\begin{aligned} \widetilde{\mathcal{R}} &= P_+^m \widetilde{\mathcal{R}}_e P_+^m \\ &= R + \widetilde{L}^T P_+^m \mathcal{L}_e P_+^m + P_+^m \mathcal{L}_e^* P_+^m \widetilde{L} \\ &= R + \widetilde{L}^T \mathcal{L} + \mathcal{L}^* \widetilde{L} = P_+^m \mathcal{R}_e P_+^m, \end{aligned}$$

because of (2.45) combined with $\mathcal{R}_e = \widetilde{\mathcal{R}}_e$ (for the last equality, see the same argument as in the proof of (1) in Proposition 2.6).

(2) Using (1c) of Proposition 2.4 one obtains with (2.43) and (2.47)

$$\begin{aligned}\langle u, \mathcal{R}u\rangle_{\mathrm{L}_+^{2,m}} &= \left\langle u, \widetilde{\mathcal{R}}u\right\rangle_{\mathrm{L}_+^{2,m}} + x^T(0)\widetilde{X}x(0) \\ &= \left\langle u, \widetilde{\mathcal{R}}u\right\rangle_{\mathrm{L}_+^{2,m}} + \left\langle u, \widehat{\Psi}^*\widetilde{X}\widehat{\Psi}u\right\rangle_{\mathrm{L}_+^{2,m}} \qquad \forall u \in \mathrm{L}_+^{2,m}\end{aligned}$$

from which the conclusion follows. □

Remark 2.4 Unlike (2) of Proposition 2.6, (2) of Proposition 2.7 shows that, in this case $\mathcal{R}$ is not really a Toeplitz operator. To be more specific let $Q = -C^TC$. In this case $\widetilde{X}$ is exactly the anticausal observability Gramian of the triple (A, B, C). Hence $\widetilde{X} = \widehat{\Theta}^*\widehat{\Theta}$ where $\widehat{\Theta}$ is the anticausal observability operator, that is,

$$\left(\widehat{\Theta}\xi\right)(t) = e^{At}\xi \qquad t \leq 0. \tag{2.48}$$

Hence

$$\mathcal{R} = \widetilde{\mathcal{R}} + \widehat{\Psi}^*\widehat{\Theta}^*\widehat{\Theta}\widehat{\Psi} = \widetilde{\mathcal{R}} + \widehat{\mathbf{H}}^*\widehat{\mathbf{H}}, \tag{2.49}$$

where $\widehat{\mathbf{H}} := \widehat{\Theta}\widehat{\Psi}$ is the anticausal Hankel operator associated with the triple (A, B, C). Therefore, we conclude from (2.49) that $\mathcal{R}$ has a mixed structure, namely a Toeplitz plus Hankel structure. This conclusion is typical for the anticausal case and introduces a kind of asymmetry with respect to the preceeding case. □

Proposition 2.8 *Let $\Sigma = (A, B; Q, L, R)$ be a Popov triplet with A antistable. Let $\widetilde{\Sigma} = \left(A, B; 0, \widetilde{L}, R\right)$ and $\Sigma_d = \left(-A^T, -\widetilde{L}; 0, B, R\right)$ be the reduced equivalent and the dual of Σ, respectively. If $\mathcal{P}_d$ and $\mathcal{R}_d$ are the operators (2.36) and (2.37) associated with Σ_d then:*

(1)

$$\mathcal{R}_d = \widetilde{\mathcal{R}};$$

(2)

$$\mathcal{P}_d = -\widehat{\Psi},$$

where $\widetilde{\mathcal{R}}$ is associated with $\widetilde{\Sigma}$.

Proof. (1) The initial value problem (2.1) is now

$$\dot{x} = -A^Tx - \widetilde{L}u \qquad x(0) = \xi.$$

Hence

$$\left(\Phi_d\xi\right)(t) = e^{-A^Tt}\xi, \qquad t \geq 0,$$

and

$$(\mathcal{L}_d u)(t) = -\int_0^t e^{-A^T(t-\tau)}\widetilde{L}u(\tau)d\tau \qquad \forall u \in \mathrm{L}_+^{2,m}.$$

Then

$$\Phi_d^* x = \int_0^\infty e^{-At}x(t)dt \qquad \forall x \in \mathrm{L}_+^{2,n} \tag{2.50}$$

and

$$(\mathcal{L}_d^* x)(t) = -\int_t^\infty \widetilde{L}^T e^{A(t-\tau)}x(\tau)d\tau \qquad \forall x \in \mathrm{L}_+^{2,n}. \tag{2.51}$$

Hence

$$\mathcal{L}_d^* B = \widetilde{L}^T \mathcal{L}, \tag{2.52}$$

and (2.52) yields

$$\mathcal{R}_d = R + B^T\mathcal{L}_d + \mathcal{L}_d^* B = R + \mathcal{L}^*\widetilde{L} + \widetilde{L}^T\mathcal{L} = \widetilde{\mathcal{R}}.$$

Similarly, (2.50) yields

$$\mathcal{P}_d u = \Phi_d^* B u = \int_0^\infty e^{-At}Bu(t)dt = -\widehat{\Psi}u \qquad \forall u \in \mathrm{L}_+^{2,m}.$$

□

2.3. THE STABILIZING SOLUTION

The next theorem is the crucial result of this chapter.

Theorem 2.1 *Let $\Sigma = (A, B; Q, L, R)$ be a Popov triplet with A stable. Then the following two statements are equivalent:*

(1) The operator $\mathcal{R}$ defined through (2.37) is boundedly invertible;

(2) R is nonsingular and the KPYS(Σ, J) (2.9) with $J = \mathrm{sgn}R$ has a stabilizing solution (X, V, W).

If (1) holds then X is given by the representation formula

$$X = \widetilde{X} - \mathcal{P}\mathcal{R}^{-1}\mathcal{P}^*, \tag{2.53}$$

where $\widetilde{X}$ is the unique solution to (2.41).

Proof. (1)⇒(2) First of all notice that $\mathcal{R}_e$ defined by (2.30) is bounded invertible because of the bounded invertibility of its Toeplitz operator $\mathcal{R}$ (see (2) of Proposition 2.6). As Π_Σ is the symbol of $\mathcal{R}_e$, it follows that $\Pi_\Sigma(j\omega)$ is bounded invertible on $\mathrm{L}^{2,m}$. Hence Π_Σ is biproper, and consequently $R = \Pi_\Sigma(\pm\infty)$ is invertible. Further, in the light of (3) of Proposition 2.4

combined with Proposition 2.6, we shall work with the reduced equivalent $\widetilde{\Sigma} = \left(A, B; 0, \widetilde{L}, R\right)$ of Σ. Following (2.8), $\Pi_{\widetilde{\Sigma}}(s)$ is realized by

$$\begin{aligned} \dot{x} &= Ax && + Bu \\ \dot{\lambda} &= && -A^T\lambda - \widetilde{L}u \\ v &= \widetilde{L}^T x + B^T\lambda + Ru. \end{aligned} \tag{2.54}$$

See now (2.54) as an initial value problem for $(\xi, u) \in \mathbf{R}^n \times \mathrm{L}_+^{2,m}$. Then the first and the second equation in (2.54) has the unique solutions

$$x^{(\xi,u)} = \Phi\xi + \mathcal{L}u \tag{2.55}$$

and

$$\lambda^u(t) = \int_t^\infty e^{-A^T(t-\tau)}\widetilde{L}u(\tau)d\tau, \quad t \geq 0, \tag{2.56}$$

respectively. Thus (2.56) yields

$$B^T\lambda^u(t) = \int_t^\infty B^T e^{-A^T(t-\tau)}\widetilde{L}u(\tau)d\tau = \left(\mathcal{L}^*\widetilde{L}u\right)(t), \quad t \geq 0, \tag{2.57}$$

as follows from (2.39). Using (2.55) and (2.57) in the last equation (2.54) one obtains

$$\begin{aligned} v &= v^{(\xi,u)} \\ &= \widetilde{L}^T x^{(\xi,u)} + B^T\lambda^u + Ru \\ &= \widetilde{L}^T\Phi\xi + \left(\widetilde{L}^T\mathcal{L} + \mathcal{L}^*\widetilde{L} + R\right)u \\ &= \widetilde{\mathcal{P}}^*\xi + \widetilde{\mathcal{R}}u = \mathcal{P}^*\xi + \mathcal{R}u, \end{aligned} \tag{2.58}$$

as follows from (2.36) and (2.37) and where Proposition 2.6 has been invoked. Since $\mathcal{R}$ is boundedly invertible, (2.58) shows that for each $\xi \in \mathbf{R}^n$, there exists a *unique* $u \in \mathrm{L}_+^{2,m}$ say u^ξ, which zeros the output v in (2.54), *i.e.* $v^{(\xi,u^\xi)} = 0$. Such a u is given explicitly by

$$u^\xi = -\mathcal{R}^{-1}\mathcal{P}^*\xi. \tag{2.59}$$

as (2.58) reveals. For $u = u^\xi$ given by (2.59), denote by x^ξ and λ^ξ the corresponding solutions $x^{(\xi,u)}$ and λ^u given by (2.55) and (2.56), respectively. Then one can write

$$0 = \widetilde{L}^T x^\xi + B^T\lambda^\xi + Ru^\xi. \tag{2.60}$$

Notice now that equality (2.60) holds in $\mathrm{L}_+^{2,m}$. Hence it is true pointwise almost everywhere. Since R is nonsingular and both x^ξ and λ^ξ are absolutely continuous on $[0, \infty)$, (2.60) reveals that for

$$u^\xi = -R^{-1}\widetilde{L}^T x^\xi - R^{-1}B^T\lambda^\xi \tag{2.61}$$

u^ξ can be chosen absolutely continuous in $L_+^{2,m}$ and (2.60) becomes a pointwise equality in $[0,\infty)$.

Using now (2.56) one deduces that

$$\begin{aligned}\lambda^\xi(0) &= \int_0^\infty e^{A^T\tau}\widetilde{L}u^\xi(\tau)d\tau \\ &= \Phi^*\widetilde{L}u^\xi \\ &= -\mathcal{P}\mathcal{R}^{-1}\mathcal{P}^*\xi. \end{aligned} \tag{2.62}$$

Let

$$\widehat{X} := -\mathcal{P}\mathcal{R}^{-1}\mathcal{P}^* = \widehat{X}^T \in \mathbf{R}^{n\times n}. \tag{2.63}$$

Then (2.62) reads

$$\lambda^\xi(0) = \widehat{X}\xi = \widehat{X}x^\xi(0). \tag{2.64}$$

For any $\varphi : [0,\infty) \to R^n$ let

$$\left(\sigma^\theta\varphi\right)(t) = \varphi(t+\theta)$$

for any $\theta \geq 0$ and $t \in [0,\infty)$. Since u^ξ, x^ξ and λ^ξ, with $x^\xi(0) = \xi$, zero the output v of the time invariant system (2.54), the same is also true for $\sigma^\theta u^\xi$, $\sigma^\theta x^\xi$ and $\sigma^\theta\lambda^\xi$ with $\left(\sigma^\theta x^\xi\right)(0) = x^\xi(\theta)$ and any $\theta \geq 0$. Owing to the uniqueness of the zeroing input u with respect to the initial condition (see (2.59)), (2.64) is updated as

$$\left(\sigma^\theta\lambda^\xi\right)(0) = \widehat{X}\left(\sigma^\theta x^\xi\right)(0),$$

or equivalently $\lambda^\xi(\theta) = \widehat{X}x^\xi(\theta)$ for all $\theta \geq 0$. Thus we conclude that

$$\lambda^\xi(t) = \widehat{X}x^\xi(t) \quad \forall t \geq 0 \tag{2.65}$$

for $\widehat{X}$ given by (2.62). With (2.65), (2.61) becomes

$$u^\xi(t) = Fx^\xi(t), \tag{2.66}$$

where

$$F = -R^{-1}\left(\widetilde{L}^T + B^T\widehat{X}\right). \tag{2.67}$$

Substituting (2.65) and (2.66) in (2.54) one gets for $v = 0$

$$\begin{aligned}\dot{x}^\xi(t) &= (A+BF)x^\xi(t), \qquad x^\xi(0) = \xi, \\ \widehat{X}\dot{x}^\xi(t) &= \left(-A^T\widehat{X} - \widetilde{L}F\right)x^\xi(t), \\ 0 &= \left(L^T + B^T\widehat{X} + RF\right)x^\xi(t),\end{aligned}$$

for $t \geq 0$. Using now the first equation in the second, one obtains at $t = 0$

$$\begin{aligned} \left(\widehat{X}(A+BF) + A^T\widehat{X} + \widetilde{L}F\right)\xi &= 0, \\ \left(\widetilde{L}^T + B^T\widehat{X} + RF\right)\xi &= 0. \end{aligned}$$

Since ξ is arbitrary the above equations yield

$$\begin{aligned} A^T\widehat{X} + \widehat{X}A + \left(\widetilde{L} + \widehat{X}B\right)F &= 0, \\ \widetilde{L}^T + B^T\widehat{X} + RF &= 0. \end{aligned} \tag{2.68}$$

Let

$$X := \widetilde{X} + \widehat{X}. \tag{2.69}$$

Since $\widetilde{L} = L + \widetilde{X}B$, (2.68) becomes with (2.69)

$$\begin{aligned} A^TX + XA + Q + (L + XB)F &= 0, \\ L + B^TX + RF &= 0. \end{aligned} \tag{2.70}$$

Write

$$R = V^TJV \tag{2.71}$$

where $J = \operatorname{sgn} R$ and where clearly V is nonsingular. Define, further,

$$W := -VF. \tag{2.72}$$

With (2.71) and (2.72) one can trivially show that the KPYS(Σ, J) (2.9) is fulfilled for (X, V, W).

Let us show finally that F is a stabilizing feedback gain, *i.e.*, $A + BF$ is stable.

From (2.55) one obtains

$$x^\xi = \left(\Phi - \mathcal{L}\mathcal{R}^{-1}\mathcal{P}^*\right)\xi.$$

Hence

$$\left\|x^\xi\right\|_2 \leq \rho \left\|\xi\right\|_2 \tag{2.73}$$

for $\rho := \left\|\Phi - \mathcal{L}\mathcal{R}^{-1}\mathcal{P}^*\right\|$. But

$$x^\xi(t) = e^{(A+BF)\,t}\xi,$$

and (2.73) yields

$$\begin{aligned} 0 &< P := \int_0^\infty e^{(A+BF)^T t} e^{(A+BF)\,t} dt \\ &\leq \rho I. \end{aligned}$$

For the positive definite matrix P defined above it is a matter of simple computation to check that

$$(A+BF)^T P + P(A+BF) + I = 0.$$

Hence by standard Lyapunov stability arguments the stability of $A+BF$ follows. Thus the implication (1)⇒(2) is completely proved.

(2)⇒(1) In this case, according to Remark 2.1 the ARE(Σ) has a stabilizing solution X and, implicitly (2.11) holds. Consequently (2.22) holds too where $S_F(s)$ is a unit in $\mathrm{RH}^\infty_{+,m\times m}$ since both A and $A+BF$ are stable. The time domain version of (2.22) reads

$$\mathcal{R}_e = \mathcal{S}^* R \mathcal{S}, \tag{2.74}$$

where S_F is the symbol of $\mathcal{S}$. Now (2.74) reveals that the Toeplitz operator $\mathcal{R}$ associated with $\mathcal{R}_e$ is

$$\mathcal{R} = \mathcal{T}^* R \mathcal{T}. \tag{2.75}$$

Here $\mathcal{T} := P_+^m \mathcal{S} P_+^m = \mathcal{T} P_+^m$ and it is explicitly defined by

$$\begin{aligned} \dot{x} &= Ax + Bu, \ \ x(0) = 0 \\ v &= -Fx + u. \end{aligned} \tag{2.76}$$

Since $A+BF$ is stable it follows that

$$\begin{aligned} \dot{x} &= (A+BF)x + Bv, \quad x(0) = 0 \\ u &= Fx + v \end{aligned}$$

is a bounded inverse of $\mathcal{T}$. Since $\mathcal{T}$ is boundedly invertible it follows from (2.75) that $\mathcal{R}$ is boundedly invertible as well and implication (2)⇒(1) is completely proved.

Finally, if we substitute (2.63) in (2.69), the representation formula (2.53) is obtained and the proof of the theorem ends. □

Corollary 2.1 *Let $\Sigma = (A, B; Q, L, R)$ be a Popov triplet with A stable. Then the following two statements are equivalent:*

(1) $\mathcal{R}$ is boundedly invertible.

(2) R is nonsingular and Π_Σ is J-factorizable, that is, there exists a unit $G \in \mathrm{RH}^\infty_{+,m\times m}$ such that

$$\Pi_\Sigma(s) = G^*(s) J G(s), \tag{2.77}$$

with $J = \operatorname{sgn} R$ and

$$G := \left[\begin{array}{c|c} A & B \\ \hline W & V \end{array}\right] \tag{2.78}$$

and where (2.77) is called a J-factorization of Π_Σ.

Proof. (1)⇒(2) Using the implication (1)⇒(2) of Theorem 2.1 it follows that the KPYS(Σ, J) ($J = \text{sgn}R$ with R nonsingular) has a stabilizing solution (X, V, W). Moreover the spectral factorization identity (2.22) holds. Using (2.71) and (2.72), (2.78) is trivially derived from (2.22). As $F = -V^{-1}W$ is a stabilizing feedback clearly (2.78) is a unit in $\text{RH}^{\infty}_{+,m\times m}$.

(2)⇒(1) The proof runs similarly to that given for the implication (2)⇒(1) of Theorem 2.1. □

Corollary 2.2 *Let $\Sigma = (A, B; Q, L, R)$ be a Popov triplet with A stable. Then the following two statements are equivalent:*

(1)

$$\Pi_{\Sigma}(j\omega) > 0 \quad \forall \omega \in [-\infty, \infty] ;$$

(2) $R > 0$ and the KPYS(Σ, I_m)

$$\begin{aligned} R &= V^T V, \\ L + XB &= W^T V, \\ Q + A^T X + XA &= W^T W, \end{aligned} \tag{2.79}$$

has a stabilizing solution (X, V, W).

The system (2.79) is usually termed as the standard Kalman–Popov–Yakubovich system, *or the* KPYS, in the positivity form.

Proof. If (1) holds then $R = \Pi_{\Sigma}(\pm\infty) > 0$ and $\text{sgn}R = I_m$. Furthermore $\mathcal{R}_e \gg 0$ as follows from the Parseval identity. Consequently $\mathcal{R} \gg 0$, i.e., $\mathcal{R}$ is boundedly invertible, and (1)⇒(2) is proved. For (2)⇒(1) use the same path as in the proof of implication (2)⇒(1) of Theorem 2.1. □

Corollary 2.3 (Bounded Real Lemma) *Let*

$$T = \left[\begin{array}{c|c} A & B \\ \hline C & D \end{array} \right].$$

Then the following two statements are equivalent:

(1) A is stable and $\|T\|_{\infty} < \gamma$

(2) For $\Sigma_c = \left(A, B; C^T C, C^T D, -\gamma^2 I + D^T D\right)$ the KPYS$(\Sigma_e, -I_m)$ has a stabilizing solution (X, V, W) with $X \geq 0$.

Proof. (1)⇒(2) The statement (1) is equivalent to $-\gamma^2 I + T^*(j\omega)T(j\omega) < 0$ for $\omega \in [-\infty, \infty]$ and $\Pi_{\Sigma_c} = -\gamma^2 I + T^* T$ as can be easily verified. Applying now (1)⇒(2) of Corollary 2.2 with the inequality '>' changed into '<' and the KPYS(Σ, I_m) replaced by KPYS$(\Sigma, -I_m)$, the conclusion follows.

(2)⇒(1) The last equation in (2.79) reads in the present context as

$$A^T X + XA + C^T C + W^T W = 0.$$

Since $X \geq 0$ and the pair $\left(\begin{bmatrix} C \\ W \end{bmatrix}, A\right)$ is detectable because of the stability of

$$A + \begin{bmatrix} 0 & -BV^{-1} \end{bmatrix} \begin{bmatrix} C \\ W \end{bmatrix} = A - BV^{-1}W,$$

it follows from standard Lyapunov stability results that A is stable. Further, to complete the proof, invoke implication (2)⇒(1) of Theorem 2.1. □

Corollary 2.4 *Let*

$$T = \left[\begin{array}{c|c} A & B \\ \hline C & D \end{array}\right],$$

with the pair (A, B) *stabilizable,* $D^T D > 0$ *and*

$$\text{rank} \begin{bmatrix} j\omega I - A & -B \\ C & D \end{bmatrix} = n + m \qquad \forall \omega \in \mathbf{R}. \tag{2.80}$$

Then the KPYS(Σ_p, I_m*), where* $\Sigma_p = \left(A, B; C^T C, C^T D, D^T D\right)$, *has a stabilizing solution* (X, V, W) *with* $X \geq 0$.

Proof. Let $\widetilde{F}$ be such that $\widetilde{A} = A + B\widetilde{F}$ is stable. Let also $\widetilde{C} := C + D\widetilde{F}$. Then (2.80) is equivalent to

$$\text{rank} \begin{bmatrix} j\omega I - \widetilde{A} & -B \\ \widetilde{C} & D \end{bmatrix} = n + m \qquad \forall \omega \in \mathbf{R}, \tag{2.81}$$

which yields

$$\widetilde{T}^*(j\omega)\widetilde{T}(j\omega) > 0 \qquad \forall \omega \in [-\infty, \infty],$$

where

$$\widetilde{T} = \left[\begin{array}{c|c} \widetilde{A} & B \\ \hline \widetilde{C} & D \end{array}\right].$$

But $\Pi_{\widetilde{\Sigma}_p} = \widetilde{T}^*\widetilde{T}$, where $\widetilde{\Sigma}_p = \left(A, B; \widetilde{C}^T\widetilde{C}, \widetilde{C}^T D, D^T D\right)$ is exactly the $\widetilde{F}$-equivalent of Σ_p. We can now conclude that all the hypotheses of Corollary 2.2 are fulfilled. Hence the KPYS($\widetilde{\Sigma}_p, I_m$) has a stabilizing solution $\left(X, \widetilde{V}, \widetilde{W}\right)$. Following (3) of Proposition 2.4 one concludes that the original KPYS(Σ_p, I_m) has a stabilizing solution (X, V, W). Let $F = -V^{-1}W$ be the stabilizing feedback gain. Then by invoking again (3) of Proposition 2.4 the last equation of the KPYS(Σ_p, I_m) can be rewritten as

$$(A + BF)^T X + X(A + BF) + (C + DF)^T (C + DF) = 0.$$

As $A + BF$ is stable it follows that $X \geq 0$. □

Corollary 2.5 *Let (A, B, C) be a linear system with the pairs (A, B) and (C, A) stabilizable and detectable, respectively. Then for*

$$\Sigma_s = \left(A, B; C^T C, 0, I\right)$$

the KPYS(Σ_s, I_m) has a stabilizing solution (X, V, W) with $X \geq 0$ or equivalently, the ARE(Σ_s)

$$A^T X + XA - XBB^T X + C^T C = 0 \tag{2.82}$$

has a positive semidefinite stabilizing solution X.

Proof. Take

$$T = \left[\begin{array}{c|c} A & B \\ \hline C & 0 \\ 0 & I \end{array}\right] = \left[\begin{array}{c|c} A & B \\ \hline C_0 & D_0 \end{array}\right] \tag{2.83}$$

and notice that the pair (A, B) is stabilizable, $D_0^T D_0 = I > 0$, and

$$\begin{aligned} \operatorname{rank}\begin{bmatrix} j\omega I - A & -B \\ C_0 & D_0 \end{bmatrix} &= \operatorname{rank}\begin{bmatrix} j\omega I - A & -B \\ C & 0 \\ 0 & I \end{bmatrix} \\ &= \operatorname{rank}\begin{bmatrix} j\omega I - A & 0 \\ C & 0 \\ 0 & I \end{bmatrix} \\ &= m + \operatorname{rank}\begin{bmatrix} j\omega I - A \\ C \end{bmatrix} \\ &= m + n \quad \forall \omega \in \mathbf{R}. \end{aligned}$$

Therefore all assumptions made in Corollary 2.4 hold. Consequently this corollary works with respect to (2.83), and the conclusion follows. □

Let us now relax the stability assumption made on A in Corollary 2.2, up to the dichotomy assumption. Then we have

Corollary 2.6 *Let $\Sigma = (A, B; Q, L, R)$ be a Popov triplet with A dichotomous and the pair (A, B) stabilizable.*

(1) $\Pi_\Sigma(j\omega) > 0 \;\; \forall \omega \in [-\infty, \infty]$;

(2) $R > 0$ and the KPYS(Σ, I_m) (2.79) has a stabilizing solution (X, V, W).

Proof. Assume that (1) holds and let $\widetilde{F}$ be such that $A + B\widetilde{F}$ is stable. Let $\widetilde{\Sigma} = \left(\widetilde{A}, B; \widetilde{Q}, \widetilde{L}, R\right)$ be the $\left(0, \widetilde{F}\right)$-equivalent of Σ. Then according to (2) of Proposition 2.4 one obtains

$$\Pi_\Sigma = S_{\widetilde{F}}^* \Pi_{\widetilde{\Sigma}} S_{\widetilde{F}}, \tag{2.84}$$

where

$$S_{\widetilde{F}} = \left[\begin{array}{c|c} A & B \\ \hline -\widetilde{F} & I \end{array}\right], \; S_{\widetilde{F}}^{-1} = \left[\begin{array}{c|c} \widetilde{A} & B \\ \hline \widetilde{F} & I \end{array}\right].$$

As A is dichotomous and $\widetilde{A}$ is stable it follows that $S_{\widetilde{F}}$ is a unit in $RL^{\infty}_{m\times m}$. Hence (2.84) reveals that $\Pi_{\widetilde{\Sigma}}(j\omega) > 0 \;\; \forall\omega \in [-\infty, \infty]$ and thus we are in the conditions of Corollary 2.2. Therefore the KPYS($\widetilde{\Sigma}, I_m$) has a stabilizing solution. Invoking now (3) of Proposition 2.4 we deduce immediately the existence of the stabilizing solution to the original KPYS(Σ, I_m). Thus the implication (1)⇒(2) is proved. For the implication (2)⇒(1) use the same path as in Corollary 2.1. □

Corollary 2.7 (Bounded Real Lemma: The relaxed version)
Let

$$T = \left[\begin{array}{c|c} A & B \\ \hline C & D \end{array}\right], \qquad y = Tu,$$

with A dichotomous and the pair (A, B) stabilizable. Then the following statements are equivalent:

(1) $\|T\|_\infty < \gamma$;

(2) For $\Sigma_c = \left(A, B; C^TC, C^TD, -\gamma^2 I + D^TD\right)$ the KPYS($\Sigma_c, -I_m$) has a stabilizing solution (X, V, W).

Proof. (1)⇒(2) Since $\Pi_{\Sigma_c} = -\gamma^2 I + T^*T$ and condition (1) in the statement holds, the conclusion follows from Corollary 2.6.

(2)⇒(1) Assume that the KPYS($\Sigma_c, -I_m$) has a stabilizing solution (X, V, W). As A is dichotomous the extended Popov index $J_{\Sigma,e}$ is well defined and it can be expressed as

$$J_{\Sigma,e}(u) = -\left\|Vu + Wx\right\|_2^2 \tag{2.85}$$

as directly follows from (2.17). Hence

$$J_{\Sigma,e}(u) \le 0. \tag{2.86}$$

But

$$J_{\Sigma,e}(u) = -\gamma^2 \left\|u\right\|_2^2 + \left\|y\right\|_2^2. \tag{2.87}$$

By combining (2.86) and (2.87) one obtains

$$\frac{\|y\|_2}{\|u\|_2} \le \gamma.$$

Hence $\|T\|_\infty \le \gamma$. Let us show that the inequality is strict. If $\|T\|_\infty = \gamma$ there exists a sequence $(u_k)_{k\ge 0}$, $u_k \in \mathrm{L}^{2,m}$ with $\|u_k\|_2 = 1$ such that

$\|y_k\|_2 \to \gamma$ as $k \to \infty$, where u_k and y_k are related via the system T, that is

$$\begin{aligned} \dot{x}_k &= Ax_k + Bu_k, \\ y_k &= Cx_k + Du_k. \end{aligned} \tag{2.88}$$

Therefore, in accordance with (2.85) and (2.87) we deduce that

$$\begin{aligned} J_{\Sigma,e}(u_k) &= -\|Vu_k + Wx_k\|_2^2 \\ &= -\gamma^2 \|u_k\|_2^2 + \|y_k\|_2^2 \\ &\to 0, \text{ as } k \to \infty. \end{aligned} \tag{2.89}$$

Since (X, V, W) is a stabilizing solution, V is nonsingular and $F = -V^{-1}W$ makes $A + BF$ stable. Then (2.88) yields

$$\begin{aligned} \dot{x}_k &= (A + BF)x_k + B(u_k - Fx_k), \\ y_k &= (C + DF)x_k + D(u_k - Fx_k). \end{aligned} \tag{2.90}$$

As $(A + BF)$ is stable and $u_k - Fx_k \in \mathrm{L}^{2,m}$, there exists $\rho > 0$ such that

$$\|y_k\|_2 \le \rho \|u_k - Fx_k\|_2 . \tag{2.91}$$

Using now (2.89) we deduce that $\|u_k - Fx_k\|_2 \to 0$ as $k \to \infty$ and (2.91) yields $\|y_k\|_2 \to 0$. But this conclusion contradicts $\|y_k\|_2 \to \gamma$ which we have just deduced above. Thus $\|T\|_\infty < \gamma$, and the proof ends. □

Let us finish this section with the 'antistable' version of Theorem 2.1.

Theorem 2.2 *Let $\Sigma = (A, B; Q, L, R)$ be a Popov triplet with A antistable and let $\widetilde{\mathcal{R}}$ be the Toeplitz operator associated with $\mathcal{R}_e$, i.e., related to the reduced equivalent $\widetilde{\Sigma} = \left(A, B; 0, \widetilde{L}, R\right)$ of Σ. If $\Sigma_d = \left(-A^T, -\widetilde{L}; 0, B, R\right)$ is the dual of Σ then the following two assertions are equivalent:*

(1) $\widetilde{\mathcal{R}}$ has a bounded inverse;

(2) R is nonsingular and the KPYS(Σ_d, J)

$$\begin{aligned} R &= \widehat{V}^T J \widehat{V}, \\ B - Z\widetilde{L} &= \widehat{W}^T J \widehat{V}, \\ -AZ - ZA^T &= \widehat{W}^T J \widehat{W}, \end{aligned} \tag{2.92}$$

with $J = \operatorname{sgn} R$, has a stabilizing solution $\left(Z, \widehat{V}, \widehat{W}\right)$.

If (1) holds then the following representation formula

$$Z = -\widehat{\Psi}\widetilde{\mathcal{R}}^{-1}\widehat{\Psi}^*$$

is true.

Proof. The proof follows trivially by combining Theorem 2.1 with Proposition 2.8. □

Remark 2.5 Corollaries 2.1 and 2.2 can be easily adapted to the situation stated in Theorem 2.2. □

We shall finish this section with the so called Small Gain Theorem which in fact is a simple corollary of the Bounded Real Lemma (see Corollary 2.3)

Corollary 2.8 (Small Gain) *Let*

$$T_i = \left[\begin{array}{c|c} A_i & B_i \\ \hline C_i & D_i \end{array}\right], \quad y_i = T_i u_i, \; i = 1, 2,$$

be two exponentially stable systems, where T_1 and T_2 are $p \times m$ and $m \times p$, respectively. If $\|T_1\|_\infty < \epsilon$ and $\|T_2\|_\infty < 1/\epsilon$ for some $\epsilon > 0$ then the resultant system obtained by the closed loop connection of T_1 and T_2, that is $u_1 = y_2$, $u_2 = y_1$, is exponentially stable (see Figure 2.1).

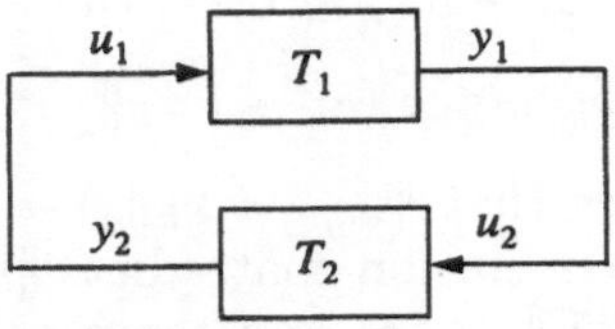

Figure 2.1.

Proof. Assume for the sake of brevity that $D_1 = 0$. By an appropriate scaling assume also that $\epsilon = 1$. Then the conditions in the statement become $\|T_i\|_\infty < 1, i = 1, 2$. Hence by invoking Corollary 2.3 the following two KPYSs in the negativeness form

$$\begin{aligned} X_1 B &= -W_1^T, \\ C_1^T C_1 + A_1^T X_1 + X_1 A_1 &= -W_1^T W_1, \end{aligned}$$

and

$$\begin{aligned} -I + D_2^T D_2 &= -V_2^T V_2, \\ C_2^T D_2 + X_2 B_2 &= -W_2^T V_2, \\ C_2^T C_2 + A_2^T X_2 + X_2 A_2 &= -W_2^T W_2, \end{aligned}$$

admit stabilizing solutions (X_1, I, W_1) and (X_2, V_2, W_2), respectively where $X_i \geq 0$, $i = 1, 2$. Let

$$x = \begin{bmatrix} x_1 \\ x_2 \end{bmatrix},$$

where x_i is the internal state of T_i, $i = 1,2$. Then the system shown in Figure 2.1 has the state space description $\dot{x} = Ax$, where

$$A = \begin{bmatrix} A_1 + B_1 D_2 C_1 & B_1 C_2 \\ B_2 C_1 & A_2 \end{bmatrix}.$$

Let

$$X = \begin{bmatrix} X_1 & 0 \\ 0 & X_2 \end{bmatrix} \geq 0.$$

Then using the above two KPYSs one can easily deduce that the following Lyapunov equation

$$A^T X + XA + C^T C = 0$$

is fulfilled for

$$C = \begin{bmatrix} W_1 + D_2 C_1 & C_2 \\ V_2 C_1 & W_2 \end{bmatrix}.$$

Since for

$$K = \begin{bmatrix} -B_1 & 0 \\ 0 & -B_2 V_2^{-1} \end{bmatrix}$$

one obtains

$$A + KC = \begin{bmatrix} A_1 - B_1 W_1 & 0 \\ 0 & A_2 - B_2 V_2^{-1} W_2 \end{bmatrix},$$

which is clearly exponentially stable, it follows that the pair (C, A) is detectable. As $X \geq 0$ the exponential stability of A follows from standard stability arguments. □

2.4. A MATRIX PENCIL DESCRIPTION

In this section we develop a KPY theory in terms of the EHP(Σ). First of all some prerequisites concerning the field of regular matrix pencils will be in order.

Recall that the matrix pencil $\lambda M - N$ where M and N are square matrices is called *regular* if $\det(\lambda M - N) \neq 0$. The *finite* spectrum of $\lambda M - N$ denoted $\Lambda_f(M, N)$ is defined as

$$\Lambda_f(M, N) = \{\lambda : \det(\lambda M - N) = 0\},$$

where λ is called a *finite generalized eigenvalue*, while the *infinite spectrum* consists of all $1/\lambda$ for which $\lambda = 0$ is a root of $\det(M - \lambda N) = 0$. In this case $1/\lambda$ is called an *infinite generalized eigenvalue.* The union of the finite and the infinite spectrum is called the *spectrum* of the pencil.

Two pencils $\lambda M - N$ and $\lambda\widetilde{M} - \widetilde{N}$ are called *strictly equivalent*, and we denote it by $\lambda M - N \sim \lambda\widetilde{M} - \widetilde{N}$, if there exist two nonsingular matrices U and W such that $\lambda\widetilde{M} - \widetilde{N} = U(\lambda M - N)W$. Clearly, if $\lambda M - N$ and $\lambda\widetilde{M} - \widetilde{N}$ are strictly equivalent they share the same spectrum finite and infinite. A deeper insight into the structure of a regular pencil is given by the following theorem:

Theorem 2.3 (Weierstrass) *Any matrix pencil $\lambda M - N$ is strictly equivalent to a matrix pencil $\lambda\widetilde{M} - \widetilde{N}$ of Weierstrass canonical form, that is,*

$$\lambda\widetilde{M} - \widetilde{N} = \begin{bmatrix} \lambda I_{n_f} - J & 0 \\ 0 & \lambda E - I_{n_\infty} \end{bmatrix}, \tag{2.93}$$

where J is in Jordan canonical form and E is in nilpotent canonical form, i.e.,

$$J = \operatorname{diag}(J_i)_{i=1}^{r}, \quad E = \operatorname{diag}\left(J_j^0\right)_{j=1}^{s},$$

where

$$J_i = \begin{bmatrix} \lambda_i & 1 & & \\ & \ddots & \ddots & \\ & & \ddots & 1 \\ & & & \lambda_i \end{bmatrix} \in \mathbf{R}^{r_i \times r_i}, \; J_j^0 = \begin{bmatrix} 0 & 1 & & \\ & \ddots & \ddots & \\ & & \ddots & 1 \\ & & & 0 \end{bmatrix} \in \mathbf{R}^{s_j \times s_j}.$$

□

A simple inspection of (2.93) shows that $\Lambda_f(M,N) = \Lambda(J)$ containing n_f finite generalized eigenvalues and the infinite spectrum has exactly $n_\infty = \sum_{j=1}^{s} s_j$ infinite generalized eigenvalues where s_j, $j = 1, \ldots, s$ are the *multiplicities* of each *infinite* generalized eigenvalue. If $s_1 = \ldots = s_s = 1$ the infinite structure of the pencil is called simple and (2.93) reduces to

$$\lambda\widetilde{M} - \widetilde{N} = \begin{bmatrix} \lambda I - J & 0 \\ 0 & -I \end{bmatrix}. \tag{2.94}$$

Definition 2.4 *Let $\lambda M - N$ ($M, N \in \mathbf{R}^{n \times n}$) be a regular pencil. A subspace $\mathcal{V} \subset \mathbf{R}^n$ of dimension ρ is called a* deflating subspace *(of the pencil) if for any basis matrix $V \in \mathbf{R}^{n \times \rho}$ of $\mathcal{V}$ there exists $S \in \mathbf{R}^{\rho \times \rho}$ such that*

$$NV = MVS. \tag{2.95}$$

If S is stable (antistable) the deflating subspace is called stable (antistable). □

The next result enlightens the significance of the deflating subspace and, in fact, justifies its name.

Proposition 2.9 *Let $\lambda M - N$ be a regular pencil. Then the following two assertions are equivalent:*

(1) $\mathcal{V}$ is a deflating subspace of dimension ρ;

(2) There exists $S \in \mathbf{R}^{\rho\times\rho}$ such that

$$\lambda M - N \sim \begin{bmatrix} \lambda I_\rho - S & \vdots & \lambda M_{12} - N_{12} \\ \cdots & \cdots & \cdots \\ 0 & \vdots & \lambda M_{22} - N_{22} \end{bmatrix}. \tag{2.96}$$

Proof. (1)$\Rightarrow$(2) Let $V \in \mathbf{R}^{n\times\rho}$ be a basis matrix for $\mathcal{V}$. Then MV is of full rank, *i.e.*, $\operatorname{rank} MV = \rho$. Indeed if $\operatorname{rank} MV = \rho' < \rho$ there exists a nonsingular matrix U' such that

$$U'MV = \begin{bmatrix} M_{11} \\ 0 \\ \cdots \\ 0 \end{bmatrix} \begin{matrix} \}\rho' \\ \}\rho-\rho' \\ \\ \}n-\rho \end{matrix} \tag{2.97}$$

where M_{11} is of full row rank ρ'. Let $\widetilde{V}$ be any completion such that $W = \begin{bmatrix} V & \widetilde{V} \end{bmatrix}$ is nonsingular. Then

$$\begin{aligned} & U'(\lambda M - N)W \\ = \; & U'\begin{bmatrix} \lambda MV - NV & \vdots \; \lambda M\widetilde{V} - N\widetilde{V} \end{bmatrix} \\ = \; & U'\begin{bmatrix} MV(\lambda I - S) \; \vdots \; \lambda M\widetilde{V} - N\widetilde{V} \end{bmatrix} \\ = \; & \begin{bmatrix} U'MV(\lambda I - S) & \vdots \; \lambda M\widetilde{V} - N\widetilde{V} \end{bmatrix} \end{aligned}$$

$$= \begin{bmatrix} \overbrace{M_{11}(\lambda I - S)}^{\rho} & \vdots & \\ & \vdots & \lambda M_{12}' - N_{12}' \\ 0 & \vdots & \\ \cdots & \cdots & \cdots \\ 0 & \vdots & \lambda M_{22}' - N_{22}' \end{bmatrix} \begin{matrix} \}\rho' \\ \\ \Big\}\rho-\rho' \\ \\ \}n-\rho \end{matrix} \tag{2.98}$$

Taking into account the last equality in (2.98) it follows that $\det U'(\lambda M - N)W = 0$, a contradiction. Hence $\rho' = \rho$ and we can write (2.98) simply as

$$U'(\lambda M - N)W = \begin{bmatrix} M_{11}(\lambda I - S) & \vdots & \lambda M'_{12} - N'_{12} \\ \cdots & \cdots & \cdots \\ 0 & \vdots & \lambda M'_{22} - N'_{22} \end{bmatrix}$$

with M_{11} $\rho \times \rho$ nonsingular. Let

$$U = \begin{bmatrix} M_{11}^{-1} & 0 \\ 0 & I_{n-\rho} \end{bmatrix} U'$$

and (2.96) follows.

(2)⇒(1) Let the nonsingular matrices U and W be those for which

$$UMW = \begin{bmatrix} I_\rho & \vdots & M_{12} \\ \cdots & \cdots & \cdots \\ 0 & \vdots & M_{22} \end{bmatrix}, \quad UNW = \begin{bmatrix} S & \vdots & N_{12} \\ \cdots & \cdots & \cdots \\ 0 & \vdots & N_{22} \end{bmatrix},$$

and let $W = \begin{bmatrix} V & \widetilde{V} \end{bmatrix}$ be partitioned conformably. Then $UNV = UMVS$ and the conclusion follows. □

In the light of (2.96) we shall write

$$\Lambda_f(M, N) \mid \mathcal{V} = \Lambda(S)$$

for the restriction of the finite spectrum of the pencil to $\mathcal{V}$. Write now J in (2.93) as

$$J = \begin{bmatrix} J_- & & \\ & J_0 & \\ & & J_+ \end{bmatrix},$$

where $J_+ \in \mathbf{R}^{n_+ \times n_+}$, $J_0 \in \mathbf{R}^{n_0 \times n_0}$, $J_- \in \mathbf{R}^{n_- \times n_-}$, $n_f = n_+ + n_- + n_0$ with $\Lambda(J_-) \subset \mathbf{C}^-$, $\Lambda(J_0) \subset \mathbf{C}^0$ and $\Lambda(J_+) \subset \mathbf{C}^+$.

Theorem 2.4 *Let $\lambda M - N$ be regular. Then it has a unique maximal stable (antistable) subspace of dimension n_- (n_+).*

Proof. Assume from the beginning that $\lambda M - N$ is brought into Weierstrass canonical form, *i.e.*,

$$\lambda \widetilde{M} - \widetilde{N} = \begin{bmatrix} \lambda I_{n_-} - J_- & & & \\ & \lambda I_{n_0} - J_0 & & \\ & & \lambda I_{n_+} - J_+ & \\ & & & \lambda E - I_{n_\infty} \end{bmatrix}. \tag{2.99}$$

Let

$$\widetilde{V}_- = \begin{bmatrix} I_{n_-} \\ 0 \\ 0 \\ 0 \end{bmatrix} \begin{matrix} \\ \}n_0 \\ \}n_+ \\ \}n_\infty \end{matrix} .$$

Then (2.99) reveals that

$$\widetilde{N}\widetilde{V}_- = \begin{bmatrix} J_- \\ 0 \\ 0 \\ 0 \end{bmatrix} = \begin{bmatrix} I_{n_-} \\ 0 \\ 0 \\ 0 \end{bmatrix} J_- = \widetilde{M}\widetilde{V}_- J_-,$$

which proves the assertion in the statement. For the parenthesized text take

$$\widetilde{V}_+ = \begin{bmatrix} 0 \\ 0 \\ I_{n_+} \\ 0 \end{bmatrix} \begin{matrix} \}n_- \\ \}n_0 \\ \\ \}n_\infty \end{matrix} .$$

□

Let $\Sigma = (A, B; Q, L, R)$ be any Popov triplet and let us explore the basic properties of the associated EHP(Σ) that will be fully exploited for developing an appropriate KPY theory in matrix pencil terms.

The next result gives a complete characterization of the structure of the EHP(Σ).

Theorem 2.5 *If the EHP(Σ) is regular then the following hold:*

(1)

$$n_- = n_+ ;$$

(2)

$$rank_{\mathbf{R}(\lambda)}\,(\lambda M - N) = 2n + m,$$

where

$$rank_{\mathbf{R}(\lambda)}\,\Pi_\Sigma(\lambda) = m;$$

(3)

$$n_\infty = m + \pi_0 ,$$

where π_0 is the number of infinite zeros of $\Pi_\Sigma(\lambda)$.

Proof. (1) For

$$P = \begin{bmatrix} 0 & -I_n & 0 \\ I_n & 0 & 0 \\ 0 & 0 & I_m \end{bmatrix}$$

one obtains

$$P^T(\lambda M - N)^T P = -\lambda M - N. \tag{2.100}$$

Hence, since the EHP is regular ($\det(\lambda M - N) \neq 0$) (2.100) shows that λ is a finite generalized eigenvalue of the EHP if and only if $-\lambda$ is a finite generalized eigenvalue of the EHP with the same multiplicity. Based on this conclusion the equality $n_+ = n_-$ follows trivially.

(2) As $\lambda M - N$ is the transmission matrix of $\Pi_\Sigma(\lambda)$ as (2.8) shows and $\lambda M - N$ is regular the conclusion follows trivially.

(3) By direct checking one obtains

$$\det(M - \lambda N) = (-1)^m \lambda^m \det(I - \lambda A) \det\left(I + \lambda A^T\right) \det \Pi_\Sigma\left(\frac{1}{\lambda}\right).$$

Hence the number of zeros in the origin of $\det(M - \lambda N)$ equals m plus the number of the zeros in the origin of $\Pi_\Sigma\left(\frac{1}{\lambda}\right)$, *i.e.*, the zeros at infinity of $\Pi_\Sigma(\lambda)$. Therefore $n_\infty = m + \pi_0$. □

Definition 2.5 *The EHP(Σ) is called* dichotomous *if $n_0 = 0$ and $n_\infty = m$ or equivalently, $n_0 = 0$ and $\pi_0 = 0$.* □

Remark 2.6 A necessary condition for the regular EHP(Σ) to be dichotomous is the invertibility of the matrix R. Indeed, since $\Pi_\Sigma(\infty) = R$ and the condition $\pi_0 = 0$ is required, the invertibility of R follows. □

The following result gives a characterization of the dichotomy in terms of deflating subspaces.

Theorem 2.6 *A regular EHP(Σ) is dichotomous if and only if it has a stable (antistable) deflating subspace of dimension n.*

Proof. Since the EHP(Σ) is regular, (2.99) shows that

$$n_- + n_0 + n_+ + n_\infty = 2n + m. \tag{2.101}$$

If the EHP is dichotomous then $n_0 = 0$, $n_\infty = m$ and (2.101) gives $n_- + n_+ = 2n$. Following (1) of Theorem 2.5 one deduces that $n_- = n_+ = n$. Further, in the light of Theorem 2.4 the EHP must have a stable (antistable) deflating subspace of dimension n.

Conversely, if the EHP(Σ) has a stable (antistable) deflating subspace of dimension n then according to the maximality of the stable (antistable) deflating subspace stated in Theorem 2.4, we conclude that

$$n_- = n_+ \geq n. \tag{2.102}$$

But (3) of Theorem 2.5 says that $n_\infty = \pi_0 + m$. Using this equality in (2.101) one obtains

$$n_- + n_+ + n_0 + \pi_0 = 2n. \tag{2.103}$$

By combining (2.102) with (2.103) one gets $2n + n_0 + \pi_0 \leq 2n$, *i.e.*, $n_0 + \pi_0 = 0$. Hence $n_0 = \pi_0 = 0$ and the dichotomy follows in accordance with Definition 2.5. □

Remark 2.7 Theorem 2.6 is equivalent to the fact that EHP(Σ) is dichotomous if and only if

$$NV = MVS, \tag{2.104}$$

where

$$V = \begin{bmatrix} V_1 \\ V_2 \\ V_3 \end{bmatrix} \begin{matrix} \}n \\ \}n \\ \}m \end{matrix} \tag{2.105}$$

is a basis matrix for the n-dimensional stable (antistable) deflating subspace and S is $n \times n$ stable (antistable). □

Definition 2.6 *The EHP(Σ) is called* stable (antistable) disconjugate *if it is dichotomous and for any basis matrix (2.105) of the n-dimensional stable (antistable) deflating subspace V_1 is $n \times n$ nonsingular.* □

Clearly, the disconjugacy is independent of the particular choice of the basis (2.105).

Finally, we shall express the so called *neutral* property of a stable (antistable) deflating subspace not necessarily maximal.

Proposition 2.10 *Let $V \in \mathbf{R}^{(2n+m)\times p}$, $p \leq n$ be a basis matrix as in (2.105) of a stable (antistable) deflating subspace of the regular EHP(Σ). Then*

$$V_1^T V_2 = V_2^T V_1. \tag{2.106}$$

Equality (2.106) expresses the so called neutral property.

Proof. Writing explicitly (2.95) one obtains:

$$\begin{aligned} AV_1 \qquad\quad + BV_3 &= V_1 S, \\ -QV_1 - A^T V_2 - LV_3 &= V_2 S, \\ L^T V_1 + B^T V_2 + RV_3 &= 0\,, \end{aligned} \tag{2.107}$$

from which the first two equations give

$$\begin{aligned} S^T V_1^T V_2 &= V_1^T A^T V_2 + V_3^T B^T V_2\,, \\ V_1^T V_2 S &= -V_1^T Q V_1 - V_1^T A^T V_2 - V_1^T L V_3\,, \\ &= -V_1^T Q V_1 - V_1^T A^T V_2 + V_2^T B V_3 + V_3^T R V_3\,, \end{aligned} \tag{2.108}$$

where the last equation in (2.107) has also been used. By adding the two equations (2.108) one obtains

$$S^T V_1^T V_2 + V_1^T V_2 S + \widehat{Q} = 0, \tag{2.109}$$

where

$$\widehat{Q} = V_1^T Q V_1 - V_3^T B^T V_2 - V_2^T B V_3 - V_3^T R V_3 = \widehat{Q}^T.$$

As S is stable (antistable), (2.109) has a unique symmetric solution which coincides with $V_1^T V_2$. Thus $V_1^T V_2 = V_2^T V_1$ and the proof ends. □

Now we are able to state and prove the main result of this section.

Theorem 2.7 *Let $\Sigma = (A, B; Q, L, R)$ be a Popov triplet. Then the following two assertions are equivalent:*

(1) R is nonsingular and the KPYS(Σ, J) (2.9), with $J = \operatorname{sgn} R$, has a (anti-) stabilizing solution (X, V, W).

(2) The EHP(Σ) is regular and (anti-) stable disconjugate.

Furthermore, if (2) holds and

$$V_0 = \begin{bmatrix} V_1 \\ V_2 \\ V_3 \end{bmatrix} \begin{matrix} \}n \\ \}n \\ \}m \end{matrix}$$

is any basis matrix for the n-dimensional (anti-) stable deflating subspace of the EHP(Σ) then

$$X = V_2 V_1^{-1}, \tag{2.110}$$

and

$$F = V_3 V_1^{-1} \tag{2.111}$$

is the (anti-) stabilizing feedback gain. Further, $W = -VF$ for any J-factorization $R = V^T J V$ of R.

Proof. (1)⇒(2) Since R is nonsingular $\Pi_\Sigma(\lambda)$ is biproper. Hence (2.8) reveals that the EHP(Σ) is regular. Further, for

$$A + BF =: S \tag{2.112}$$

write the KPYS(Σ, J) (2.9) as

$$-Q - A^T X - XA = W^T J W = W^T J V F = (L + XB)F,$$

or equivalently,

$$-Q - A^T X - LF = XS. \tag{2.113}$$

Further,

$$L^T + B^T X = V^T J W = -RF,$$

or equivalently,

$$L^T + B^T X + RF = 0. \tag{2.114}$$

An aggregated form of (2.112), (2.113) and (2.114) is

$$\begin{bmatrix} A & 0 & B \\ -Q & -A^T & -L \\ L^T & B^T & R \end{bmatrix} \begin{bmatrix} I \\ X \\ F \end{bmatrix} = \begin{bmatrix} I & 0 & 0 \\ 0 & I & 0 \\ 0 & 0 & 0 \end{bmatrix} \begin{bmatrix} I \\ X \\ F \end{bmatrix} S,$$

or

$$NV_0 = V_0 MS, \tag{2.115}$$

for

$$V_0 = \begin{bmatrix} I \\ X \\ F \end{bmatrix}.$$

As S in (2.115) is $n \times n$ (anti-) stable it follows from Theorem 2.6 that the EHP(Σ) is dichotomous. Since in addition the $n \times n$ submatrix of V_0 placed in the top of it is the identity matrix, the disconjugacy follows.

(2)⇒(1) We can write (2.107) with V_1 nonsingular and S $n \times n$ (anti-) stable. Using (2.110) and (2.111), (2.107) gives

$$\begin{aligned} A + BF = V_1 S V_1^{-1} &= \widehat{S}, \\ -Q - A^T X - LF &= X\widehat{S}, \\ L^T + B^T X + RF &= 0, \end{aligned} \tag{2.116}$$

where $\widehat{S}$ is $n \times n$ (anti-) stable.

In addition, since V_0 is neutral, as Proposition 2.10 asserts, it follows, by using the disconjugacy, that

$$V_1^{-T} V_2^T = V_2 V_1^{-1},$$

that is, $X = X^T$. With this conclusion (2.114) reduces to the KPYS(Σ, J) (2.9) for $F = -V^{-1}W$ and V given by any J-factorization $R = V^T J V$ of R. Thus the proof is completed. □

2.5. THE SIGNATURE CONDITION

As we have seen in Theorem 2.1, a necessary and sufficient condition for the existence of a stabilizing solution to the KPYS is the bounded invertibility of the Toeplitz operator $\mathcal{R}$. Such a condition is not easily checkable. However there are two relevant cases where this condition can be easily checked. The first case is the Popov positiveness condition mentioned in Corollary 2.2. A second case will be now under attention. It worthwhile stressing that this condition is the keypoint for solving the tasks of modern control synthesis.

Theorem 2.8 *Let $\Sigma = (A, B; Q, L, R)$ be a Popov triplet with B partitioned as*

$$B = [\ \overbrace{B_1}^{m_1}\ \ \overbrace{B_2}^{m_2}\ ] \qquad (m = m_1 + m_2).$$

Let L, R and Π_Σ be partitioned conformably, i.e.,

$$\begin{array}{c} L = [\ \overbrace{L_1}^{m_1}\ \ \overbrace{L_2}^{m_2}\ ]\ , \\ R = \begin{bmatrix} \overbrace{R_{11}}^{m_1} & \overbrace{R_{12}}^{m_2} \\ R_{12}^T & R_{22} \end{bmatrix}, \quad \Pi_\Sigma = \begin{bmatrix} \Pi_{11} & \Pi_{12} \\ \Pi_{12}^* & \Pi_{22} \end{bmatrix} \end{array} \tag{2.117}$$

and let $\Sigma_2 = (A, B_2; Q, L_2, R_{22})$. If the following three assumptions:

(a) A is stable;

(b) $\Pi_{22} > 0$ on $j\overline{\mathbf{R}}$;

(c) There exists $S \in \mathrm{RH}^\infty_{+,m_2 \times m_1}$ such that

$$[\ I \quad S^*\ ]\,\Pi_\Sigma \begin{bmatrix} I \\ S \end{bmatrix} < 0 \quad on \ \ j\overline{\mathbf{R}}; \tag{2.118}$$

all hold, then the following are true:

(1) $R_{22} > 0$ and the KPYS(Σ, I_{m_2}) has a stabilizing solution (X_2, V_2, W_{22});

(2)

$$\operatorname{sgn} R = J = \begin{bmatrix} -I_{m_1} & \\ & I_{m_2} \end{bmatrix} \tag{2.119}$$

and the KPYS(Σ, J) has a stabilizing solution (X, V, W) with V of lower left block-triangular form

$$V = \begin{bmatrix} V_{11} & 0 \\ V_{21} & V_{22} \end{bmatrix} \tag{2.120}$$

where the partition of V is taken in accordance with R in (2.117) and, in addition,

$$X \geq X_2. \tag{2.121}$$

Furthermore, if

$$\begin{bmatrix} Q & L_2 \\ L_2^T & R_{22} \end{bmatrix} \geq 0 \tag{2.122}$$

then

$$X \geq 0. \tag{2.123}$$

Condition (2.118) is called the signature condition.

Proof. (1) follows trivially from Corollary 2.2.

(2) Let $\mathcal{R}_e : \mathrm{L}^{2,m} \to \mathrm{L}^{2,m}$ and $\mathcal{S}_e : \mathrm{L}^{2,m_1} \to \mathrm{L}^{2,m_2}$ be the time domain operators associated with Π_Σ and S, respectively. As $S \in \mathrm{RH}^\infty_{+,m_2\times m_1}$ clearly the following equality holds for the Toeplitz operator $\mathcal{S}$ associated with $\mathcal{S}_e$

$$\mathcal{S} = P_+^{m_2}\mathcal{S}_e P_+^{m_1} = \mathcal{S}_e P_+^{m_1}. \tag{2.124}$$

Let $\mathcal{R}_e$ and the Toeplitz operator $\mathcal{R} = P_+^m \mathcal{R}_e P_+^m$ associated with it, be partitioned in accordance with R in (2.117), *i.e.*

$$\mathcal{R}_e = \begin{bmatrix} \mathcal{R}_{e11} & \mathcal{R}_{e12} \\ \mathcal{R}_{e12}^* & \mathcal{R}_{e22} \end{bmatrix}$$

and

$$\mathcal{R} = \begin{bmatrix} \mathcal{R}_{11} & \mathcal{R}_{12} \\ \mathcal{R}_{12}^* & \mathcal{R}_{22} \end{bmatrix},$$

respectively. The time domain counterpart of (2.118) reads

$$\begin{bmatrix} I & \mathcal{S}_e^* \end{bmatrix} \begin{bmatrix} \mathcal{R}_{e11} & \mathcal{R}_{e12} \\ \mathcal{R}_{e12}^* & \mathcal{R}_{e22} \end{bmatrix} \begin{bmatrix} I \\ \mathcal{S}_e \end{bmatrix} =: \mathcal{T}_{e11} \ll 0. \tag{2.125}$$

Let $\mathcal{T}_{11} = P_{m_1}^+ \mathcal{T}_{e11} P_{m_1}^+$. Clearly $\mathcal{T}_{11} \ll 0$.
Using (2.124), (2.125) yields

$$\begin{aligned} 0 &\gg P_{m_1}^+ \mathcal{T}_{e11} P_{m_1}^+ \\ &= \mathcal{T}_{11} \\ &= P_{m_1}^+ \begin{bmatrix} I & \mathcal{S}_e^* \end{bmatrix} \begin{bmatrix} \mathcal{R}_{e11} & \mathcal{R}_{e12} \\ \mathcal{R}_{e12}^* & \mathcal{R}_{e22} \end{bmatrix} \begin{bmatrix} I \\ \mathcal{S}_e \end{bmatrix} P_{m_1}^+ \\ &= \begin{bmatrix} I & \mathcal{S}^* \end{bmatrix} \begin{bmatrix} P_{m_1}^+ & \\ & P_{m_2}^+ \end{bmatrix} \begin{bmatrix} \mathcal{R}_{e11} & \mathcal{R}_{e12} \\ \mathcal{R}_{e12}^* & \mathcal{R}_{e22} \end{bmatrix} \\ &\quad \times \begin{bmatrix} P_{m_1}^+ & \\ & P_{m_2}^+ \end{bmatrix} \begin{bmatrix} I \\ \mathcal{S} \end{bmatrix} \\ &= \begin{bmatrix} I & \mathcal{S}^* \end{bmatrix} \begin{bmatrix} \mathcal{R}_{11} & \mathcal{R}_{12} \\ \mathcal{R}_{12}^* & \mathcal{R}_{22} \end{bmatrix} \begin{bmatrix} I \\ \mathcal{S} \end{bmatrix} \\ &= \mathcal{R}_{11} + \mathcal{R}_{12}\mathcal{S} + \mathcal{S}^*\mathcal{R}_{12}^* + \mathcal{S}^*\mathcal{R}_{22}\mathcal{S}. \end{aligned} \tag{2.126}$$

Using (2.126) and taking into account that $\mathcal{R}_{22} \gg 0$, one obtains

$$\begin{aligned} 0 &\gg \mathcal{T}_{11} - (\mathcal{R}_{12} + \mathcal{S}^*\mathcal{R}_{22})\,\mathcal{R}_{22}^{-1}\,(\mathcal{R}_{12}^* + \mathcal{R}_{22}\mathcal{S}) \\ &= \mathcal{R}_{11} + \mathcal{R}_{12}\mathcal{S} + \mathcal{S}^*\mathcal{R}_{12} + \mathcal{S}^*\mathcal{R}_{22}\mathcal{S} - \mathcal{R}_{12}\mathcal{R}_{22}^{-1}\mathcal{R}_{12}^* - \mathcal{R}_{12}\mathcal{S} \\ &\quad -\mathcal{S}^*\mathcal{R}_{12} - \mathcal{S}^*\mathcal{R}_{22}\mathcal{S} = \mathcal{R}_{11} - \mathcal{R}_{12}\mathcal{R}_{22}^{-1}\mathcal{R}_{12}^* =: \mathcal{R}_{11}^{\times} \end{aligned} \tag{2.127}$$

where $\mathcal{R}_{11}^{\times}$ is the Schur complement of $\mathcal{R}_{22}$ in $\mathcal{R}$. Hence $\mathcal{R}_{11}^{\times} \ll 0$, as follows from (2.127), and in addition $\mathcal{R}_{22} \gg 0$, as follows from condition (b) in the statement. These two conclusions lead undoubtedly to the bounded invertibility of $\mathcal{R}$. Hence according to Theorem 2.1 $\mathcal{R}$ has a bounded inverse and the KPYS(Σ, J), with $J = \operatorname{sgn} R$, has a stabilizing solution (X, V, W). We shall show that J coincides exactly with that shown in (2.119) and V has the structure indicated in (2.120). From (b) it follows that

$$R_{22} = \Pi_{22}(j\infty) > 0 \tag{2.128}$$

and from (c) one obtains for $\omega = \infty$

$$\begin{bmatrix} I & S_\infty^T \end{bmatrix} \begin{bmatrix} R_{11} & R_{12} \\ R_{12}^T & R_{22} \end{bmatrix} \begin{bmatrix} I \\ S_\infty \end{bmatrix} < 0, \tag{2.129}$$

where $S_\infty := S(j\infty)$. By combining (2.128) with (2.129) in the same manner as we proceeded for the signature of $\mathcal{R}$ above, one obtains

$$R_{11}^{\times} = R_{11} - R_{12}R_{22}^{-1}R_{12}^T < 0. \tag{2.130}$$

From (2.128) and (2.129) one deduces that

$$\begin{aligned} &\begin{bmatrix} R_{11} & R_{12} \\ R_{12}^T & R_{22} \end{bmatrix} \\ &= \begin{bmatrix} I & R_{12}R_{22}^{-1} \\ 0 & I \end{bmatrix} \begin{bmatrix} R_{11}^{\times} & 0 \\ 0 & R_{22} \end{bmatrix} \begin{bmatrix} I & 0 \\ R_{22}^{-1}R_{12}^T & I \end{bmatrix}. \end{aligned} \tag{2.131}$$

Thus (2.131) yields simultaneously that

$$\operatorname{sgn} R = \begin{bmatrix} -I_{m_1} & \\ & I_{m_2} \end{bmatrix}$$

and

$$V = \begin{bmatrix} \left(-R_{11}^{\times}\right)^{\frac{1}{2}} & 0 \\ R_{22}^{-\frac{1}{2}}R_{12}^T & R_{22}^{\frac{1}{2}} \end{bmatrix} \equiv \begin{bmatrix} V_{11} & 0 \\ V_{21} & V_{22} \end{bmatrix}.$$

Let us now prove (2.121). Using the representation formula (2.53) we have

$$\begin{aligned} X &= \widetilde{X} - \begin{bmatrix} \mathcal{P}_1 & \mathcal{P}_2 \end{bmatrix} \begin{bmatrix} \mathcal{R}_{11} & \mathcal{R}_{12} \\ \mathcal{R}_{12}^T & \mathcal{R}_{22} \end{bmatrix}^{-1} \begin{bmatrix} \mathcal{P}_1^* \\ \mathcal{P}_2^* \end{bmatrix} \\ &= \widetilde{X} - \mathcal{P}_2 \mathcal{R}_{22}^{-1} \mathcal{P}_2^* \\ &\quad - \left(\mathcal{P}_1 - \mathcal{P}_2 \mathcal{R}_{22}^{-1} \mathcal{R}_{12}^*\right) \left(\mathcal{R}_{11}^{\times}\right)^{-1} \left(\mathcal{P}_1 - \mathcal{P}_2 \mathcal{R}_{22}^{-1} \mathcal{R}_{12}^*\right)^*, \quad (2.132) \end{aligned}$$

where $\mathcal{P}$ has been partitioned conformably. As $X_2 = \widetilde{X} - \mathcal{P}_2 \mathcal{R}_{22}^{-1} \mathcal{P}_2^*$ and $\mathcal{R}_{11}^{\times} \ll 0$, (2.126) implies (2.121) automatically. Finally, if (2.122) holds, clearly $X_2 \geq 0$ and (2.123) follows from (2.121). Thus the proof is completed. □

NOTES AND REFERENCES

The KPYS and the Popov function originate in the positivity theory developed in (Popov, 1973) and (Yakubovich, 1975). More recent considerations concerning the role of the Popov function in the algebraic Riccati theory are given in (Lancaster and Rodman, 1991), (Ionescu and Weiss, 1993), (Ionescu *et al.*, 1997). The main concepts introduced in this chapter as well as an extensive treatement of the generalized Popov–Yakubovich theory are presented in (Ionescu and Weiss, 1993), (Halanay and Ionescu, 1993), (Halanay and Ionescu, 1994). Results on deflating subspace methods, matrix pencils and connections with AREs can be found in (Stewart, 1973), (Pappas *et al.*, 1980), (Van Dooren, 1981), (Van Dooren, 1983), (Ionescu and Weiss, 1992), (Varga, 1992). More recent results including the extension of deflating subspaces to the case of general pencils, possibly singular, are derived in (Oara, 1995), (Ionescu and Oara, 1996) and (Oara and Van Dooren, 1997).

For early references concerning the Bounded Real Lemma we cite (Anderson, 1967), (Willems, 1971), (Zhou and Khargonekar, 1988). More recently, a new version in a more general setting is given in (Ben-Artzi *et al.*, 1995).

A pioneering work on the Small Gain Theorem is (Zames, 1966).

CHAPTER 3

H^∞ CONTROL: A SIGNATURE CONDITION BASED APPROACH

This chapter is dedicated to a problem that undoubtedly dominated the Mathematical Control Theory during the last decade. Roughly speaking, the problem consists in finding a controller which simultaneously stabilizes and achieves disturbance attenuation under a prescribed tolerance level. In fact, the enormous attention paid to the problem above sketched can be explained by the necessity of providing appropriate solutions of the current generation of applications which have posed new kinds of control requirements. Among these the question of *robustness* is crucial. To be more specific, we have to say that, as in fact is well known, the fundamental requirement of feedback systems is to achieve stability and accomplish performance objectives not only for a single nominal model, but also for a set of models covering expected uncertainties about the model parameters and perturbations. Controllers which possesses this property are called 'robust'. What was spectacular for the H^∞-control theory was that this theory provided an elegant and complete solution of the *robust control problem.*

There are several 'traditional' ways to approach the H^∞-control theory. Unlike these, in what follows we shall 'attack' the problem via the *signature condition* combined with KPYS techniques as they were presented in the preceeding chapter.

3.1. PROBLEM STATEMENT

Let

$$\begin{aligned} T &= \left[\begin{array}{c|cc} A & B_1 & B_2 \\ \hline C_1 & D_{11} & D_{12} \\ C_2 & D_{21} & 0 \end{array}\right] \\ &= \left[\begin{array}{c|c} A & B \\ \hline C & D \end{array}\right] = \begin{bmatrix} T_{11} & T_{12} \\ T_{21} & T_{22} \end{bmatrix}, \\ \begin{bmatrix} y_1 \\ y_2 \end{bmatrix} &= T \begin{bmatrix} u_1 \\ u_2 \end{bmatrix} \end{aligned} \tag{3.1}$$

be a generalized system. Here $x \in \mathbf{R}^n$ is the state, $u_1 \in \mathbf{R}^{m_1}$ is the vector of *external inputs*, $y_1 \in \mathbf{R}^{p_1}$ is the vector of *regulated outputs*, $u_2 \in \mathbf{R}^{m_2}$ is the vector of *control inputs*, and $y_2 \in \mathbf{R}^{p_2}$ is the *measurement* vector. A, B_i, C_i, D_{ij} ($D_{22} = 0$) $i, j = 1, 2$ are matrices of appropriate dimensions. Let also be introduced the aggregated input and output vectors

$$u = \begin{bmatrix} u_1 \\ u_2 \end{bmatrix} \in \mathbf{R}^m, \quad y = \begin{bmatrix} y_1 \\ y_2 \end{bmatrix} \in \mathbf{R}^p.$$

Any linear system

$$K = \left[\begin{array}{c|c} A_K & B_K \\ \hline C_K & D_K \end{array}\right], \quad y_K = K u_K, \tag{3.2}$$

is called a *controller* for the system (3.1) if $u_K = y_2$ and $u_2 = y_K$ (see Figure 3.1)

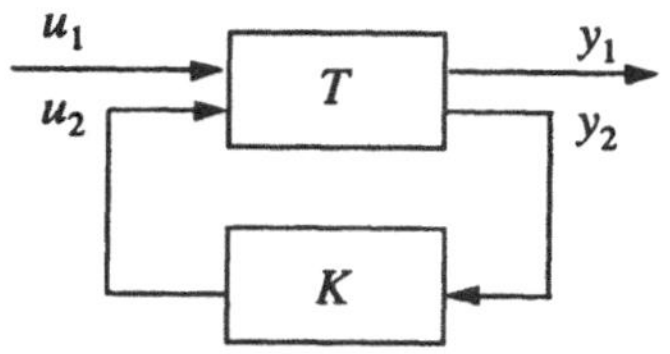

Figure 3.1.

The resultant closed loop system shown in Figure 3.1 is made explicit by

$$T_{y_1 u_1} = \left[\begin{array}{c|c} A_R & B_R \\ \hline C_R & D_R \end{array}\right], \quad y_1 = T_{y_1 u_1} u_1, \tag{3.3}$$

where

$$\begin{aligned} A_R &= \begin{bmatrix} A + B_2 D_K C_2 & B_2 C_K \\ B_K C_2 & A_K \end{bmatrix}, \\ B_R &= \begin{bmatrix} B_1 + B_2 D_k D_{21} \\ B_K D_{21} \end{bmatrix}, \\ C_R &= \begin{bmatrix} C_1 + D_{12} D_K C_2 & D_{12} C_K \end{bmatrix}, \\ D_R &= D_{11} + D_{12} D_K D_{21}. \end{aligned} \tag{3.4}$$

The H^∞-control problem can be stated as follows: for a prescribed tolerance level $\gamma > 0$ find a controller (3.2) such that the resultant closed loop system (3.3) possesses simultaneously two properties:

(P1) *stability*, that is,

$$\Lambda(A_R) \subset \mathbf{C}^-; \tag{3.5}$$

(P2) γ-*attenuation*, that is,

$$\|T_{y_1 u_1}\|_\infty < \gamma, \tag{3.6}$$

where the L^∞-norm of $T_{y_1u_1}$ is well defined because of (P1), *i.e.*, $T_{y_1u_1} \in RH^\infty_{+,p_1\times m_1}$.

The transfer matrix $T_{y_1u_1}$ can also be expressed only in terms of the partial transfer matrices T_{ij} and the transfer matrix K. Indeed, (3.1) and (3.2) gives

$$\begin{aligned} y_1 &= T_{11}u_1 + T_{12}u_2, \\ y_2 &= T_{21}u_1 + T_{22}u_2, \\ u_2 &= Ky_2. \end{aligned} \tag{3.7}$$

With the last equation in (3.7) substituted in the second, one obtains

$$(I - T_{22}K)\, y_2 = T_{21}u_1. \tag{3.8}$$

As $T_{22}(\infty)K(\infty) = 0$, $I - T_{22}K$ is biproper, and (3.8) gives

$$y_2 = (I - T_{22}K)^{-1}\, T_{21}u_1 =: T_{y_2u_1}u_1. \tag{3.9}$$

Further, the last equation of (3.7) combined with (3.9) yields

$$u_2 = K\,(I - T_{22}K)^{-1}\, T_{21}u_1 =: T_{u_2u_1}u_1. \tag{3.10}$$

With (3.10) substituted in the first equation in (3.7) one obtains eventually

$$T_{y_1u_1} = T_{11} + T_{12}K\,(I - T_{22}K)^{-1}\, T_{21}. \tag{3.11}$$

In other words $T_{y_1u_1}$ can be expressed as a *linear fractional transformation* (LFT) performed on K, and where the coefficients of the transformation are the partial transfer matrices T_{ij}. Thus

$$T_{y_1u_1} = LFT(T, K). \tag{3.12}$$

Simple computations show that the realization of $(I - T_{22}K)^{-1}$, $T_{y_2u_1}$ and $T_{u_2u_1}$, have the same state matrix which is exactly A_R, *i.e.*,

$$\begin{aligned} (I - T_{22}K)^{-1} = \left[\begin{array}{c|c} A_R & * \\ \hline * & * \end{array}\right], \quad T_{y_2u_1} = \left[\begin{array}{c|c} A_R & * \\ \hline * & * \end{array}\right], \\ T_{u_2u_1} = \left[\begin{array}{c|c} A_R & * \\ \hline * & * \end{array}\right] \end{aligned} \tag{3.13}$$

where $*$ denotes an irrelevant entry. From (3.13) we conclude that (P1) implies

$$\begin{aligned} (I - T_{22}K)^{-1} &\in RH^\infty_{+,p_2\times p_2}, \\ T_{y_2u_1} &\in RH^\infty_{+,p_2\times m_1}, \\ T_{u_2u_1} &\in RH^\infty_{+,m_2\times m_1}. \end{aligned} \tag{3.14}$$

By performing an appropriate scaling we can take

$$\gamma = 1. \tag{3.15}$$

Indeed (3.11) gives

$$\begin{aligned}\gamma^{-1}T_{y_1u_1} &= \gamma^{-\frac{1}{2}}T_{11}\gamma^{-\frac{1}{2}} \\ &+ \left(\gamma^{-\frac{1}{2}}T_{12}\gamma^{\frac{1}{2}}\right)\left(\gamma^{-1}K\right) \\ &\times \left(I - \left(\gamma^{\frac{1}{2}}T_{22}\gamma^{\frac{1}{2}}\right)\left(\gamma^{-1}K\right)\right)^{-1}\left(\gamma^{-\frac{1}{2}}T_{21}\gamma^{\frac{1}{2}}\right),\end{aligned}$$

from which the scaled system and controller, in terms of the original data, are rapidly derived to be

$$T_{scaled} = \left[\begin{array}{c|cc} A & \gamma^{-\frac{1}{2}}B_1 & \gamma^{\frac{1}{2}}B_2 \\ \hline \gamma^{-\frac{1}{2}}C_1 & \gamma^{-1}D_{11} & D_{12} \\ \gamma^{\frac{1}{2}}C_2 & D_{21} & 0 \end{array}\right] \tag{3.16}$$

and

$$K_{scaled} = \left[\begin{array}{c|c} A_K & \gamma^{-\frac{1}{2}}B_K \\ \hline \gamma^{-\frac{1}{2}}C_K & \gamma^{-1}D_K \end{array}\right]. \tag{3.17}$$

Henceforward we shall consider the original data in (3.1), (3.2) as given for an H^∞ problem with unitary tolerance level, *i.e.*, (3.15) holds.

The following regularity assumptions will be now adopted:

(R1) $\operatorname{rank} D_{12} = m_2$ & $\operatorname{rank}\begin{bmatrix} j\omega I - A & -B_2 \\ C_1 & D_{12}\end{bmatrix} = n+m_2 \ \forall\, \omega \in \mathbf{R}$;

(R2) $\operatorname{rank} D_{21} = p_2$ & $\operatorname{rank}\begin{bmatrix} j\omega I - A & -B_1 \\ C_2 & D_{21}\end{bmatrix} = n+p_2 \ \forall\, \omega \in \mathbf{R}$.

3.2. THE REDHEFFER THEOREM

In this section a special case of closed loop strict subunitary disturbance attenuation is studied. As we shall see in the next sections, the theory developed in this section enables us to solve, via reduction to a simpler problem, the original problem itself. We shall start first with the following definition:

Definition 3.1 *Let*

$$G = \left[\begin{array}{c|c} A & B \\ \hline C & D \end{array}\right]$$

be a stable system. It is called inner *if*

$$D^T D = I,$$

and there exists $X = X^T \geq 0$ such that the following system is fulfilled

$$\begin{aligned} C^T D + XB &= 0, \\ C^T C + A^T X + XA &= 0. \end{aligned} \tag{3.18}$$

□

One can easily prove that

$$G^* G = I, \tag{3.19}$$

that is, G is an *isometry.* Indeed, using (1.8), one obtains

$$\begin{aligned} G^* G &= \left[\begin{array}{c|c} -A^T & -C^T \\ \hline B^T & D^T \end{array}\right] \left[\begin{array}{c|c} A & B \\ \hline C & D \end{array}\right] \\ &= \left[\begin{array}{cc|c} A & 0 & B \\ -C^T C & -A^T & -C^T D \\ \hline D^T C & B^T & D^T D \end{array}\right]. \end{aligned} \tag{3.20}$$

Further, by performing the coordinate transformation

$$\hat{x} = \left[\begin{array}{cc} I & 0 \\ -X & I \end{array}\right] x,$$

and by taking into account Definition 3.1, one obtains eventually

$$\begin{aligned} G^* G &= \left[\begin{array}{cc|c} A & 0 & B \\ -C^T C - XA - A^T X & -A^T & -C^T D - XB \\ \hline D^T C + B^T X & B^T & I \end{array}\right] \\ &= \left[\begin{array}{cc|c} A & 0 & B \\ 0 & -A^T & 0 \\ \hline 0 & B^T & I \end{array}\right] = I \end{aligned}$$

after removing the uncontrollable and unobservable parts. Let us state and prove the main result of this section called the Redheffer Theorem.

Theorem 3.1 *Assume that the following assumptions concerning the system (3.1) all hold:*

(a) T is inner;
(b) D_{21} is nonsingular and $A - B_1 D_{21}^{-1} C_2$ is stable.

Then the following two statements are equivalent:

(1) The controller K given in (3.2) stabilizes (3.1) and $\|T_{y_1 u_1}\|_\infty < 1$;
(2) A_K in (3.2) is stable and $\|K\|_\infty < 1$.

Proof. (2)⇒(1) By applying the Bounded Real Lemma (see Corollary 2.3) to the system (3.2) it follows that the following KPYS,

$$\begin{aligned} -I + D_K^T D_K &= -V_K^T V_K, \\ C_K^T D_K + X_K B_K &= -W_K^T V_K, \\ C_K^T C_K + A_K^T X_K + X_K A_K &= -W_K^T W_K, \end{aligned} \tag{3.21}$$

has a stabilizing solution (X_K, V_K, W_K) with $X_K \geq 0$. Now take into consideration that T in (3.1) is inner and make explicit this property via Definition 3.1, *i.e.*, $D^T D = I$ and (3.18) holds for an appropriate $X \geq 0$. Let

$$X_R = \begin{bmatrix} X & 0 \\ 0 & X_K \end{bmatrix}, V_R = V_K D_{21}, W_R = \begin{bmatrix} V_K C_2 & W_K \end{bmatrix}. \tag{3.22}$$

Then for A_R, B_R, C_R, D_R given by (3.4) one can check by simple manipulations that use (3.21) and (3.18), that the following KPYS,

$$\begin{aligned} -I + D_R^T D_R &= -V_R^T V_R, \\ C_R^T D_R + X_R B_R &= -W_R^T V_R, \\ C_R^T C_R + A_R^T X_R + X_R A_R &= -W_R^T W_R, \end{aligned} \tag{3.23}$$

is fulfilled for (3.22). Moreover (X_R, V_R, W_R) given by (3.22) is a stabilizing solution to (3.23). Indeed V_R is nonsingular and

$$A_R - B_R V_R^{-1} W_R = \begin{bmatrix} A - B_1 D_{21}^{-1} C_2 & * \\ 0 & A_K - B_K V_K^{-1} W_K \end{bmatrix},$$

which is clearly a stable matrix. Hence by applying the Bounded Real Lemma it follows that A_R is stable, *i.e.*, K stabilizes T and, in addition, $\|T_{y_1 u_1}\| < 1$.

(1)⇒(2) Let us show first that $K \in \mathrm{RL}^\infty_{m_2 \times m_2}$. Using (3.7) it is trivial to check that

$$\begin{bmatrix} T_{y_1 u_1} \\ T_{y_2 u_1} \end{bmatrix} = \begin{bmatrix} T_{11} & T_{12} \\ T_{21} & T_{22} \end{bmatrix} \begin{bmatrix} I \\ T_{u_2 u_1} \end{bmatrix}. \tag{3.24}$$

If both sides of (3.24) are premultiplied by T^* then by taking into account the isometry of T one obtains

$$T_{11}^* T_{y_1 u_1} + T_{21}^* T_{y_2 u_1} = I,$$

from which

$$T_{y_2 u_1} = T_{21}^{-*} (I - T_{11}^* T_{y_1 u_1}) \tag{3.25}$$

$\left(T_{21}^{-*} := (T_{21}^{*})^{-1}\right)$. Since

$$T_{21} = \left[\begin{array}{c|c} A & B_1 \\ \hline C_2 & D_{21} \end{array}\right]$$

and

$$T_{21}^{-1} = \left[\begin{array}{c|c} A - B_1 D_{21}^{-1} C_2 & B_1 D_{21}^{-1} \\ \hline -D_{21}^{-1} C_2 & D_{21}^{-1} \end{array}\right]$$

it follows that T_{21} is a unit in $\mathrm{RH}^{\infty}_{+,p_2\times m_1}$. Hence T_{21}^{*} is a unit in $\mathrm{RH}^{\infty}_{-,m_1\times m_1}$. Further, since T is isometric we have $\|T\|_{\infty} = 1$ and consequently $\|T_{11}\|_{\infty} \leq 1$. Hence $\|T_{11}^{*} T_{y_1u_1}\|_{\infty} < 1$ and we deduce that $I - T_{11}^{*} T_{y_1u_1}$ is a unit in $\mathrm{RL}^{\infty}_{m_1\times m_1}$. By combining this conclusion with the fact that T_{21}^{*} is a unit in $\mathrm{RH}^{\infty}_{-,m_1\times m_1}$, (3.25) reveals that $T_{y_2u_1}$ is a unit in $\mathrm{RL}^{\infty}_{m_1\times m_1}$. Therefore

$$K = T_{u_2u_1} T_{y_2u_1}^{-1} \in \mathrm{RL}^{\infty}_{m_2\times p_2} \tag{3.26}$$

Moreover $\|K\|_{\infty} < 1$. Assume the contrary, *i.e.*, $\|K\|_{\infty} \geq 1$. Then there exists a sequence $\left(y_2^k\right)_{k\geq 0}$ with $y_2^k \in \mathrm{L}^{2,p_2}$ and $\left\|y_2^k\right\|_2 = 1$ such that for $\left(u_2^k\right)_{k\geq 0}$ with $u_2^k \in \mathrm{L}^{2,m_2}$ defined by

$$u_2^k(j\omega) = K(j\omega) y_2^k(j\omega)$$

we have

$$\lim_{k\to\infty} \left(\left\|u_2^k\right\|_2^2 - \left\|y_2^k\right\|_2^2\right) \geq 0 \tag{3.27}$$

Since $T_{y_2u_1}$ is a unit in $\mathrm{RL}^{\infty}_{m_1\times m_1}$, one can define the input sequence $\left(u_1^k\right)_{k\geq 0}$, $u_1^k \in \mathrm{L}^{2,m_1}$ by

$$u_1^k(j\omega) = T_{y_2u_1}^{-1}(j\omega) y_2^k(j\omega) \neq 0, \ \forall k \geq 0.$$

Let further $\left(y_1^k\right)_{k\geq 0}$, $y_1^k \in \mathrm{L}^{2,p_1}$ be defined by

$$y_1^k(j\omega) = T_{y_1u_1}(j\omega) u_1^k(j\omega).$$

Since T is isometric we must have

$$\left\|u_1^k\right\|_2^2 + \left\|u_2^k\right\|_2^2 = \left\|y_1^k\right\|_2^2 + \left\|y_2^k\right\|_2^2,$$

from which

$$\left\|y_1^k\right\|_2^2 - \left\|u_1^k\right\|_2^2 = \left\|u_2^k\right\|_2^2 - \left\|y_2^k\right\|_2^2. \tag{3.28}$$

By combining (3.28) with (3.27) one obtains

$$\lim_{k\to\infty}\left(\left\|y_1^k\right\|_2^2-\left\|u_1^k\right\|_2^2\right)\geq 0$$

which contradicts the strict contractiveness of $T_{y_1u_1}$. Thus $\|K\|_\infty < 1$. Let us show that A_K is stable. Since K stabilizes T clearly the pairs (A_K, B_K) and (C_K, A_K) must be stabilizable and detectable. Using this conclusion in conjunction with (3.26) it follows that A_K must be dichotomous. Therefore, since $\|K\|_\infty < 1$ it follows via the *relaxed version of Bounded Real Lemma* (see Corollary 2.7) that the KPYS (3.21) has a stabilizing solution (X_K, V_K, W_K). By combining again (3.18) with (3.21), (3.23) will hold for (X_R, V_R, W_R) defined by (3.22). But A_R in the last equation (3.23) is stable. Hence $X_R \geq 0$ from which $X_K \geq 0$ as follows from (3.22). By combining this conclusion with the detectability of the pair (C_K, A_K), the last equation in (3.21) reveals, via standard stability result, that A_K is stable. Thus the theorem is completely proved. □

3.3. A FIRST SET OF NECESSARY SOLVABILITY CONDITIONS

In this section we shall show how a first set of necessary solvability conditions can be easily derived as a trivial consequence of the signature condition presented in Section 2.5.

Associated with the system (3.1), let the following Popov triplet $\Sigma_c = (A, B; Q_c, L_c, R_c)$ be explicitly introduced by

$$\begin{aligned} Q_c &= C_1^T C_1, \quad L_c = C_1^T \begin{bmatrix} D_{11} & D_{12} \end{bmatrix}, \\ R_c &= \begin{bmatrix} D_{11}^T \\ D_{12}^T \end{bmatrix}\begin{bmatrix} D_{11} & D_{12} \end{bmatrix} - \begin{bmatrix} I_{m_1} & 0 \\ 0 & 0 \end{bmatrix} \\ & \quad (B = \begin{bmatrix} B_1 & B_2 \end{bmatrix}). \end{aligned} \tag{3.29}$$

It is a matter of elementary computation to see that the Popov function Π_{Σ_c} associated with Σ_c is

$$\begin{aligned} \Pi_{\Sigma_c} &= \begin{bmatrix} T_{11}^* \\ T_{12}^* \end{bmatrix}\begin{bmatrix} T_{11} & T_{12} \end{bmatrix} - \begin{bmatrix} I_{m_1} & 0 \\ 0 & 0 \end{bmatrix} \\ &= \begin{bmatrix} T_{11}^*T_{11} - I & T_{11}^*T_{12} \\ T_{12}^*T_{11} & T_{12}^*T_{12} \end{bmatrix} = \begin{bmatrix} \Pi_{\Sigma_{c,11}} & \Pi_{\Sigma_{c,12}} \\ \Pi_{\Sigma_{c,12}}^* & \Pi_{\Sigma_{c,22}} \end{bmatrix}. \end{aligned} \tag{3.30}$$

Assume now that the H^∞-control problem has a solution (3.2). Hence

$$\|T_{y_1u_1}\|_\infty < 1, \tag{3.31}$$

where $T_{y_1u_1} = LFT(T, K)$ (see (3.11), (3.12)) can be written in accordance with (3.10) as

$$T_{y_1u_1} = T_{11} + T_{12}S, \tag{3.32}$$

where

$$S := T_{u_2u_1} \in \mathrm{RH}^{\infty}_{+,m_2\times m_1}. \tag{3.33}$$

Then (3.31) is equivalent to

$$(T_{11} + T_{12}S)^* (T_{11} + T_{12}S) - I < 0 \text{ on } j\overline{\mathbf{R}}, \tag{3.34}$$

or equivalently,

$$\begin{bmatrix} I & S^* \end{bmatrix} \begin{bmatrix} T_{11}^*T_{11} - I & T_{11}^*T_{12} \\ T_{12}^*T_{11} & T_{12}^*T_{12} \end{bmatrix} \begin{bmatrix} I \\ S \end{bmatrix} < 0 \text{ on } j\overline{\mathbf{R}}. \tag{3.35}$$

By comparing (3.35) with (3.30) we deduce that (3.35) receives the condensed form

$$\begin{bmatrix} I & S^* \end{bmatrix} \Pi_{\Sigma_c} \begin{bmatrix} I \\ S \end{bmatrix} < 0 \text{ on } j\overline{\mathbf{R}}. \tag{3.36}$$

Since (R1) holds, clearly

$$\Pi_{\Sigma_c,22} = T_{12}^*T_{12} > 0 \text{ on } j\overline{\mathbf{R}}. \tag{3.37}$$

In addition

$$\begin{aligned} \begin{bmatrix} Q_c & L_{c2} \\ L_{c2}^T & R_{c22} \end{bmatrix} &= \begin{bmatrix} C_1^TC_1 & C_1^TD_{12} \\ D_{12}^TC_1 & D_{12}^TD_{12} \end{bmatrix} \\ \begin{bmatrix} C_1^T \\ D_{12}^T \end{bmatrix} \begin{bmatrix} C_1 & D_{12} \end{bmatrix} &\geq 0. \end{aligned} \tag{3.38}$$

If we assume temporarily that A in Σ_c is *stable*, *i.e.*, the system (3.1) is stable, then conditions (a), (b), (c) in Theorem 2.8 all hold, and in the light of the cited theorem the KPYS(Σ_c, J_c), where

$$J_c = \begin{bmatrix} -I_{m_1} & 0 \\ 0 & I_{m_2} \end{bmatrix}, \tag{3.39}$$

has a stabilizing solution (X, V_c, W_c) with $X \geq 0$ and V_c structured as in (2.120).

Assume now that A is not necessarily stable. Since K stabilizes T clearly the pair (A, B_2) must be stabilizable. Hence there exists $\widetilde{F}_2$ such that $A +$

$B_2\widetilde{F}_2$ is stable. Let $\widetilde{F} = \begin{bmatrix} 0 \\ \widetilde{F}_2 \end{bmatrix}$ and let $\widetilde{\Sigma}_c = \left(\widetilde{A}, B; \widetilde{Q}_c, \widetilde{L}_c, R_c\right)$ be the $\left(0, \widetilde{F}\right)$-equivalent of Σ_c. Hence by applying (2.21) one obtains

$$\begin{aligned} \widetilde{A} &= A + B\widetilde{F} = A + B_2\widetilde{F}_2, \\ \widetilde{Q}_c &= \widetilde{C}_1^T \widetilde{C}_1, \\ \widetilde{L}_c &= \widetilde{C}_1^T \begin{bmatrix} D_{11} & D_{12} \end{bmatrix}, \end{aligned} \tag{3.40}$$

where $\widetilde{C}_1 := C_1 + D_{12}\widetilde{F}_2$. With (3.40) one can show immediately that

$$\Pi_{\widetilde{\Sigma}_c} = \begin{bmatrix} \widetilde{T}_{11}^* \widetilde{T}_{11} - I & \widetilde{T}_{11}^* \widetilde{T}_{12} \\ \widetilde{T}_{12}^* \widetilde{T}_{11} & \widetilde{T}_{12}^* \widetilde{T}_{12} \end{bmatrix}, \tag{3.41}$$

where

$$\begin{bmatrix} \widetilde{T}_{11} & \widetilde{T}_{12} \end{bmatrix} = \left[\begin{array}{c|cc} \widetilde{A} & B_1 & B_2 \\ \hline \widetilde{C}_1 & D_{11} & D_{12} \end{array}\right], \tag{3.42}$$

and where $\widetilde{A}$ in (3.42) is now stable. Invoking (2) of Proposition 2.4 one can write

$$\Pi_{\Sigma_c} = S_{\widetilde{F}}^* \Pi_{\widetilde{\Sigma}_c} S_{\widetilde{F}}, \tag{3.43}$$

where

$$\begin{aligned} S_{\widetilde{F}} &= \left[\begin{array}{c|c} A & B \\ \hline -\widetilde{F} & I \end{array}\right] = \left[\begin{array}{c|cc} A & B_1 & B_2 \\ \hline 0 & I & 0 \\ -\widetilde{F}_2 & 0 & I \end{array}\right] \\ &= \begin{bmatrix} I & 0 \\ S_{\widetilde{F},21} & S_{\widetilde{F},22} \end{bmatrix}. \end{aligned}$$

Hence

$$\begin{bmatrix} I & 0 \\ S_{\widetilde{F},21} & S_{\widetilde{F},22} \end{bmatrix} \begin{bmatrix} I \\ S \end{bmatrix} = \begin{bmatrix} I \\ \widetilde{S} \end{bmatrix}, \tag{3.44}$$

where

$$\widetilde{S} := S_{\widetilde{F},21} + S_{\widetilde{F},22} S. \tag{3.45}$$

Furthermore, using (3.10) and (3.33), (3.45) reveals that

$$\widetilde{S} = S_{\widetilde{F},21} + S_{\widetilde{F},22} K \left(I - T_{22}K\right)^{-1} T_{21} = LFT\left(\widetilde{T}, K\right)$$

where

$$\widetilde{T} = \begin{bmatrix} S_{\widetilde{F},21} & S_{\widetilde{F},22} \\ T_{21} & T_{22} \end{bmatrix} = \left[\begin{array}{c|cc} A & B_1 & B_2 \\ \hline -\widetilde{F}_2 & 0 & I \\ C_2 & D_{21} & 0 \end{array}\right]$$

Taking into account the expression of A_R in (3.4) we conclude immediately that K stabilizes $\widetilde{T}$, *i.e.*, $\widetilde{S} \in \mathrm{RH}^{\infty}_{+,m_2\times m_1}$. Using (3.36) in conjunction with (3.43)-(3.45) one obtains

$$\begin{aligned} &\left[\begin{array}{cc} I & S^* \end{array}\right] \Pi_{\Sigma_c} \left[\begin{array}{c} I \\ S \end{array}\right] \\ &= \left[\begin{array}{cc} I & \widetilde{S}^* \end{array}\right] \Pi_{\widetilde{\Sigma}_c} \left[\begin{array}{c} I \\ \widetilde{S} \end{array}\right] < 0 \quad \text{on} \quad j\overline{\mathbf{R}}. \end{aligned} \tag{3.46}$$

Furthermore using (3.42) in conjunction with (R1) one obtains

$$\Pi_{\widetilde{\Sigma}_c,22} = \widetilde{T}_{12}^* \widetilde{T}_{12} > 0 \quad \text{on} \quad j\overline{\mathbf{R}}, \tag{3.47}$$

and, in addition,

$$\begin{aligned} \left[\begin{array}{cc} \widetilde{Q}_c & \widetilde{L}_{c2} \\ \widetilde{L}_{c2}^T & R_{c22} \end{array}\right] &= \left[\begin{array}{cc} \widetilde{C}_1^T \widetilde{C}_1 & \widetilde{C}_1^T D_{12} \\ D_{12}^T \widetilde{C}_1 & D_{12}^T D_{12} \end{array}\right] \\ &= \left[\begin{array}{c} \widetilde{C}_1^T \\ D_{12}^T \end{array}\right] \left[\begin{array}{cc} \widetilde{C}_1 & D_{12} \end{array}\right] \geq 0. \end{aligned} \tag{3.48}$$

From (3.46)–(3.48) we conclude that Theorem 2.8 works with respect to $\widetilde{\Sigma}_c$. Since $\widetilde{\Sigma}_c$ is the $\left(0, \widetilde{F}\right)$–equivalent to Σ_c it follows, by invoking (3) of Proposition 2.4, that the same conclusion still holds with respect to the original Popov triplet Σ_c. Resuming the above development the following result can be now stated.

Theorem 3.2 *Assume that the H^∞-control problem formulated for (3.1) has a solution (3.2). Then for Σ_c and J_c introduced by (3.29) and (3.39) the KPYS(Σ_c, J_c) has a stabilizing solution*

$$(X, V_c, W_c) = \left(X, \left[\begin{array}{cc} V_{c11} & 0 \\ V_{c12} & V_{c22} \end{array}\right], \left[\begin{array}{c} W_{c1} \\ W_{c2} \end{array}\right]\right),$$

where $X \geq 0$ and the partitions of V_c and W_c are taken in accordance with J_c in (3.39). □

Now according to (3.1) we have

$$\begin{aligned} T^T &= \left[\begin{array}{c|cc} A^T & C_1^T & C_2^T \\ \hline B_1^T & D_{11}^T & D_{21}^T \\ B_2^T & D_{12}^T & 0 \end{array}\right], \\ \left[\begin{array}{c} v_1 \\ v_2 \end{array}\right] &= \left[\begin{array}{cc} T_{11}^T & T_{21}^T \\ T_{12}^T & T_{22}^T \end{array}\right] \left[\begin{array}{c} w_1 \\ w_2 \end{array}\right] \end{aligned} \tag{3.49}$$

and for (3.49) the role of the Popov triplet Σ_c and regularity condition (R1) played for (3.1) are now replaced by $\Sigma_o = \left(A^T, C^T; Q_o, L_o, R_o\right)$ and the regularity condition (R2), and where

$$\begin{aligned} Q_o &= B_1 B_1^T,\ L_o = B_1 \begin{bmatrix} D_{11}^T & D_{21}^T \end{bmatrix}, \\ R_o &= \begin{bmatrix} D_{11} \\ D_{21} \end{bmatrix} \begin{bmatrix} D_{11}^T & D_{21}^T \end{bmatrix} - \begin{bmatrix} I_p & 0 \\ 0 & 0 \end{bmatrix}. \end{aligned} \tag{3.50}$$

Let

$$J_o = \begin{bmatrix} -I_{p_1} & 0 \\ 0 & I_{p_2} \end{bmatrix}. \tag{3.51}$$

Now it is trivial to see that K stabilizes T and achieves (3.31) if and only if K^T stabilizes T^T and achieves $\left\| T_{v_1 w_1}^T \right\|_\infty < 1$. Hence we conclude that the following result, which is the dual of that stated in Theorem 3.2, holds.

Theorem 3.3 *Assume that the H^∞-control problem formulated for (3.1) has a solution (3.2). Then for Σ_o and J_o introduced through (3.50) and (3.51), respectively the KPYS(Σ_o, J_o) has a stabilizing solution*

$$(Y, V_o, W_o) = \left(Y, \begin{bmatrix} V_{o11} & 0 \\ V_{o12} & V_{o22} \end{bmatrix}, \begin{bmatrix} W_{o1} \\ W_{o2} \end{bmatrix} \right),$$

where $Y \geq 0$ and the partitions of V_o and W_o are taken in accordance with J_o in (3.51). □

Assume now that the following normalizing conditions hold

$$\begin{aligned} D_{11} &= 0, \\ D_{12}^T \begin{bmatrix} C_1 & D_{12} \end{bmatrix} &= \begin{bmatrix} 0 & I \end{bmatrix}, \\ \begin{bmatrix} B_1 \\ D_{21} \end{bmatrix} D_{21}^T &= \begin{bmatrix} 0 \\ I \end{bmatrix}. \end{aligned} \tag{3.52}$$

If (3.52) is true the entries of Σ_c and Σ_o reduce to

$$Q_c = C_1^T C_1, \quad L_c = \begin{bmatrix} 0 & 0 \end{bmatrix}, \quad R_c = \begin{bmatrix} -I_{m_1} & 0 \\ 0 & I_{m_2} \end{bmatrix} = J_c \tag{3.53}$$

and

$$Q_o = B_1 B_1^T, \quad L_o = \begin{bmatrix} 0 & 0 \end{bmatrix}, \quad R_o = \begin{bmatrix} -I_{p_1} & 0 \\ 0 & I_{p_2} \end{bmatrix} = J_o \tag{3.54}$$

as follows from (3.29), (3.50) and (3.52) respectively. According to J_c and J_o we deduce rapidly that $V_c = I_m$ and $V_o = I_p$. Hence, according to Remark

2.1, we can work with ARE(Σ_c) and ARE(Σ_o) instead of KPYS(Σ_c, J_c) and KPYS(Σ_o, J_o) respectively. Such AREs are

$$A^T X + XA - X\left(-B_1 B_1^T + B_2 B_2^T\right) X + C_1^T C_1 = 0 \tag{3.55}$$

and

$$AY + YA^T - Y\left(-C_1^T C_1 + C_2^T C_2\right) Y + B_1 B_1^T = 0, \tag{3.56}$$

as trivially follows from (3.53) and (3.54) respectively. Thus by cumulating, in this particular case, the results stated in Theorems 3.2 and 3.3, the following result holds.

Theorem 3.4 *Assume that the H^∞-control problem formulated for (3.1) has a solution (3.2). Assume additionally that the normalizing conditions (3.52) hold. Then the AREs (3.55) and (3.56) have positive semi-definite stabilizing solutions X and Y, respectively.* □

3.4. A SECOND SET OF NECESSARY SOLVABILITY CONDITIONS

Let Σ_c be the Popov triplet introduced by (3.29). Then the extended Popov index $J_{\Sigma_{c,e}}$ associated with Σ_c is

$$J_{\Sigma_{c,e}}(u) \equiv J_{\Sigma_{c,e}}(u_1, u_2) = -\|u_1\|_2^2 + \|y_1\|_2^2 \tag{3.57}$$

whenever the rightmost term makes sense. As in the previous section *assume that there exists a solution (3.2) to the H^∞-control problem formulated for (3.1).* Then clearly $J_{\Sigma_{c,e}}$ makes sense and in addition

$$-\|u_1\|_2^2 + \|y_1\|_2^2 < 0. \tag{3.58}$$

Furthermore, in the light of Theorem 3.2 the KPYS(Σ_c, J_c) has a stabilizing solution

$$(X, V_c, W_c) = \left(X, \begin{bmatrix} V_{c11} & 0 \\ V_{c12} & V_{c22} \end{bmatrix}, \begin{bmatrix} W_{c1} \\ W_{c2} \end{bmatrix}\right) \quad (X \geq 0).$$

Hence, following (2.17), one obtains

$$\begin{aligned} J_{\Sigma_{c,e}}(u_1, u_2) &= \langle V_c u + W_c x, J_c (V_c u + W_c x)\rangle \\ &= \left\langle \begin{bmatrix} V_{c11} u_1 + W_{c1} x \\ V_{c21} u_1 + V_{c22} u_2 + W_{c2} x \end{bmatrix}, \right. \\ &\quad \begin{bmatrix} -I_{m_1} & 0 \\ 0 & I_{m_2} \end{bmatrix} \\ &\quad \left. \times \begin{bmatrix} V_{c11} u_1 + W_{c1} x \\ V_{c21} u_1 + V_{c22} u_2 + W_{c2} x \end{bmatrix} \right\rangle, \end{aligned} \tag{3.59}$$

where x is the unique L^2-solution to $\dot{x} = Ax + B_1u_1 + B_2u_2$ for arbitrary $u_1 \in L^{2,m_1}$ and the corresponding $u_2 \in L^{2,m_2}$ delivered by the controller K. Moreover, $u = -V_c^{-1}W_cx$ is a stationary point (in the Fréchet sense) for $J_{\Sigma_{c,e}}$. Let

$$\begin{aligned} \tilde{u}_1 &:= V_{c11}u_1 + W_{c1}x, \\ \tilde{y}_1 &:= V_{c21}u_1 + V_{c22}u_2 + W_{c2}x. \end{aligned} \tag{3.60}$$

Then (3.59) and (3.60) yields

$$J_{\Sigma_{c,e}}(u_1, u_2) = -\|\tilde{u}_1\|_2^2 + \|\tilde{y}_1\|_2^2, \tag{3.61}$$

and, by taking into account (3.57) and (3.61), we conclude that the following identity holds

$$-\|u_1\|_2^2 + \|y_1\|_2^2 = -\|\tilde{u}_1\|_2^2 + \|\tilde{y}_1\|_2^2. \tag{3.62}$$

Notice now that (3.62) suggests the following three conjectures:

Conjecture 1 Let

$$\begin{aligned} \dot{x} &= Ax + B_1u_1 + B_2u_2 \\ y_1 &= C_1x + D_{11}u_1 + D_{12}u_2 \\ y_2 &= C_2x + D_{21}u_1 \end{aligned} \tag{3.63}$$

be the explicit state space form of (3.1). Let the following transformation

$$\begin{aligned} u_1 &= V_{c11}^{-1}\tilde{u}_1 - V_{c11}^{-1}W_{c1}x = V_{c11}^{-1}\tilde{u}_1 + F_1x \\ \tilde{y}_1 &= V_{c21}V_{c11}^{-1}\tilde{u}_1 + V_{c22}u_2 + \left(-V_{c21}V_{c11}^{-1}W_{c1} + W_{c2}\right)x \\ &= -V_{c22}F_2x + V_{c21}V_{c11}^{-1}\tilde{u}_1 + V_{c22}u_2 \end{aligned} \tag{3.64}$$

be inspired by (3.60), where

$$\begin{aligned} F = -V_c^{-1}W_c &= \begin{bmatrix} -V_{c11}^{-1}W_{c1} \\ V_{c22}^{-1}V_{c21}V_{c11}^{-1}W_{c1} - V_{c22}^{-1}W_{c2} \end{bmatrix} \\ &= \begin{bmatrix} F_1 \\ F_2 \end{bmatrix} \end{aligned} \tag{3.65}$$

is the stabilizing feedback gain associated with the stabilizing solution to the KPYS(Σ_c, J_c). By performing (3.64) on (3.63), the following new system is obtained

$$\begin{aligned} \dot{x}_o &= (A + B_1F_1)x_o + B_1V_{c11}^{-1}\tilde{u}_1 + B_2u_2, \\ \tilde{y}_1 &= -V_{c22}F_2x_o + V_{c21}V_{c11}^{-1}\tilde{u}_1 + V_{c22}u_2, \\ y_2 &= (C_2 + D_{21}F_1)x_o + D_{21}V_{c11}^{-1}\tilde{u}_1 \end{aligned} \tag{3.66}$$

(here x has been redenoted by x_o), or equivalently,

$$T_o = \left[\begin{array}{c|cc} A+B_1F_1 & B_1V_{c11}^{-1} & B_2 \\ \hline -V_{c22}F_2 & V_{c21}V_{c11}^{-1} & V_{c22} \\ C_2+D_{21}F_1 & D_{21}V_{c11}^{-1} & 0 \end{array}\right],$$

$$\begin{bmatrix} \tilde{y}_1 \\ y_2 \end{bmatrix} = T_o \begin{bmatrix} \tilde{u}_1 \\ u_2 \end{bmatrix}. \tag{3.67}$$

Then (3.62) shows that (3.58) holds if and only if $-\|\tilde{u}_1\|_2^2 + \|\tilde{y}_1\|_2^2 < 0$. Hence the following conjecture is suggested: *K is a solution to the H^∞-control problem formulated for T (see (3.63)) if and only if K is a solution to the H^∞-control problem formulated for T_o in (3.67) (see Figure 3.2)*

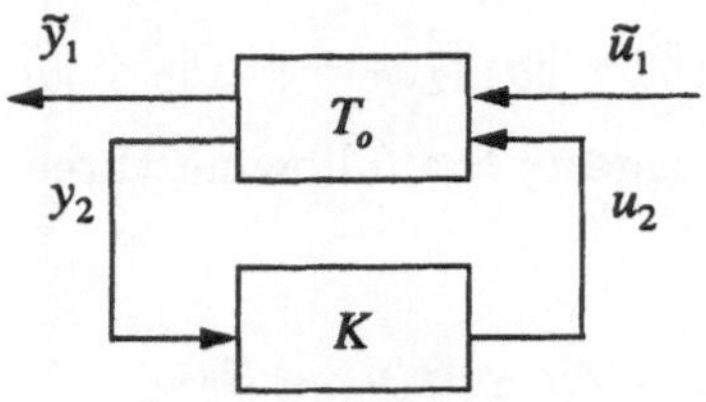

Figure 3.2.

Conjecture 2 Now read (3.62) as

$$\|u_1\|_2^2 + \|\tilde{y}_1\|_2^2 = \|y_1\|_2^2 + \|\tilde{u}_1\|_2^2 \tag{3.68}$$

and consider the new transformation

$$\begin{aligned} u_2 &= -V_{c22}^{-1}V_{c21}u_1 + V_{c22}^{-1}\tilde{y}_1 - V_{c22}^{-1}W_{c2}x \\ \tilde{u}_1 &= V_{c11}u_1 \qquad\qquad\qquad\qquad + W_{c1}x, \end{aligned} \tag{3.69}$$

also derived from (3.64). By performing (3.69) on (3.63) the following new system is obtained

$$\begin{aligned} \dot{x}_I &= \left(A - B_2V_{c22}^{-1}W_{c2}\right)x_I + \left(B_1 - B_2V_{c22}^{-1}V_{c21}\right)u_1 + B_2V_{c22}^{-1}\tilde{y}_1, \\ y_1 &= \left(C_1 - D_{12}V_{c22}^{-1}W_{c2}\right)x_I + \left(D_{11} - D_{12}V_{c22}^{-1}V_{c21}\right)u_1 \\ &\quad + D_{12}V_{c22}^{-1}\tilde{y}_1, \\ \tilde{u}_1 &= W_{c1}x_I \qquad\qquad\qquad\qquad + V_{c11}u_1 \end{aligned} \tag{3.70}$$

(here x has been redenoted by x_I), or equivalently,

$$\begin{aligned} T_I &= \left[\begin{array}{c|cc} A - B_2V_{c22}^{-1}W_{c22} & B_1 - B_2V_{c22}^{-1}V_{c21} & B_2V_{c22}^{-1} \\ \hline C_1 - D_{12}V_{c22}^{-1}W_{c2} & D_{11} - D_{12}V_{c22}^{-1}V_{c21} & D_{12}V_{c22}^{-1} \\ W_{c1} & V_{c11} & 0 \end{array}\right] \\ &= \left[\begin{array}{c|c} A_I & B_I \\ \hline C_I & D_I \end{array}\right], \quad \begin{bmatrix} y_1 \\ \tilde{u}_1 \end{bmatrix} = T_I \begin{bmatrix} u_1 \\ \tilde{y}_1 \end{bmatrix} \end{aligned} \tag{3.71}$$

(see Figure 3.3).

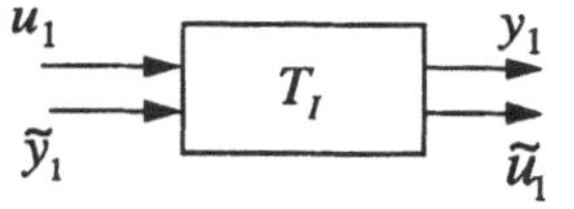

Figure 3.3.

Then (3.68) suggests the following conjecture: T_I *in (3.71) is inner.*

Conjecture 3 By coupling T_o to T_I (see Figure 3.4) the resultant system coincides, modulo stable uncontrollable/ unobservable parts, with T.

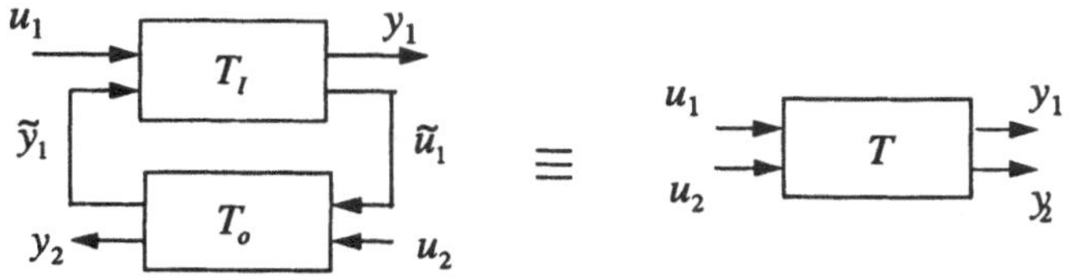

Figure 3.4.

In an appropriate terminology it says that T *is the Redheffer product of* T_I *and* T_o and it is denoted $T_I \otimes T_o$. Let us start to prove the above conjectures.

Proposition 3.1

$$T = T_I \otimes T_o \tag{3.72}$$

Proof. By coupling T_o to T_I as it is shown in the left hand side of Figure 3.4 one gets

$$T_I \otimes T_o = \left[\begin{array}{cc|cc} A + B_2F_2 & -B_2F_2 & B_1 & B_2 \\ -B_1F_2 & A + B_1F_1 & B_1 & B_2 \\ \hline C_1 + D_{12}F_2 & -D_{12}F_2 & D_{11} & D_{12} \\ C_2 + D_{21}F_1 & -D_{21}F_1 & D_{21} & 0 \end{array}\right]. \tag{3.73}$$

Performing now the coordinate change

$$\begin{bmatrix} x_I \\ \xi \end{bmatrix} = \begin{bmatrix} I & 0 \\ -I & I \end{bmatrix} \begin{bmatrix} x_I \\ x_o \end{bmatrix} = \begin{bmatrix} x_I \\ -x_I + x_o \end{bmatrix},$$

(3.73) yields

$$
\begin{aligned}
T_I \otimes T_o &= \left[\begin{array}{cc|cc} A & -B_2F_2 & B_1 & B_2 \\ 0 & A+BF & 0 & 0 \\ \hline C_1 & -D_{12}F_2 & D_{11} & D_{12} \\ C_2 & -D_{21}F_1 & D_{21} & 0 \end{array}\right] \\
&= \left[\begin{array}{c|cc} A & B_1 & B_2 \\ \hline C_1 & D_{11} & D_{12} \\ C_2 & D_{21} & 0 \end{array}\right] = T,
\end{aligned}
$$

where the stable uncontrollable part has been removed. Thus the third conjecture is proved. □

Let us now prove the second conjecture.

Proposition 3.2 *The following hold:*

(1) A_I is stable.

(2) If (X, V_c, W_c) with $X \geq 0$, is the stabilizing solution to the KPYS (Σ_c, J_c) then the following system of equations

$$
\begin{aligned}
D_I^T D_I &= I, \\
C_I^T D_I + XB_I &= 0, \\
C_I^T C_I + A_I^T X + XA_I &= 0
\end{aligned}
\tag{3.74}
$$

is fulfilled.

(3) D_{I21} is nonsingular and $A_I - B_{I1}D_{I21}^{-1}C_{I2}$ is stable.

Proof. (2) Using the explicit expressions of A_I, B_I, C_I and D_I given by (3.71), (3.74) simply follows by using the explicit form of the KPYS(Σ_c, J_c).

(1) Since for

$$K_I = \begin{bmatrix} 0 & V_{c22}^{-1}V_{c21}V_{c11}^{-1} \end{bmatrix}$$

we have $A_I + K_IC_I = A + BF$, we deduce that the pair (C_I, A_I) is detectable. By combining this conclusion with the fact that the last equation in (3.74) is fulfilled for $X \geq 0$, the stability of A_I follows from standard stability arguments.

(3) $D_{I21} = V_{c11}$ is nonsingular and $A_I - B_{I1}D_{I21}^{-1}C_{I2} = A + BF$ is stable. From (1) and (2) we conclude that T_I is inner and the second conjecture is completely proved. □

Finally, let us prove the first conjecture which in fact is the main result of this section.

Proposition 3.3 *K is a solution to H^∞-control problem formulated for T if and only if K is a solution to the H^∞-control problem formulated for T_o.*

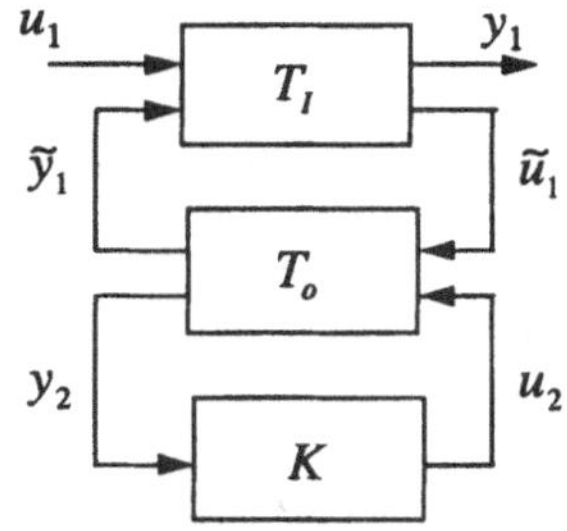

Figure 3.5.

Proof. Consider the structure shown in Figure 3.5. Assume that K is a solution to H^∞-control problem for T. Hence, according to Proposition 3.1, K is a solution to H^∞-control problem for $T_I \otimes T_o$. It follows that $T_{\tilde{y}_1\tilde{u}_1}$ is a solution to the H^∞-control problem for T_I. By combining Proposition 3.2 with the implication (2)⇒(1) of Redheffer Theorem, one deduces that $T_{\tilde{y}_1\tilde{u}_1}$ is internally stable and $\|T_{\tilde{y}_1\tilde{u}_1}\|_\infty < 1$, that is, K is a solution to H^∞-control problem for T_o. Conversely, if K is a solution to the H^∞-control problem for T_o it follows that $T_{\tilde{y}_1\tilde{u}_1}$ is internally stable and $\|T_{\tilde{y}_1\tilde{u}_1}\|_\infty < 1$. Further by invoking again Proposition 3.2 in conjunction with the implication (1)⇒(2) of Redheffer Theorem we conclude that the system structure shown in Figure 3.5 is internally stable, *i.e.*, K stabilizes $T_I \otimes T_o$, namely T, and $\|T_{y_1u_1}\|_\infty < 1$. Thus the proof is completed. □

Denote now by $\Sigma_\times = (A_\times, B_\times; Q_\times, L_\times, R_\times)$ the Popov triplet obtained by updating Σ_o, given in (3.50), with the data of T_o, *i.e.*,

$$\begin{aligned}
A_\times &= (A + B_1F_1)^T, \\
B_\times &= \begin{bmatrix} -F_2^TV_{c22}^T & (C_2 + D_{21}F_1)^T \end{bmatrix}, \\
Q_\times &= B_1\left(V_{c11}^TV_{c11}\right)^{-1}B_1^T, \\
L_\times &= B_1\left(V_{c11}^TV_{c11}\right)^{-1}\begin{bmatrix} V_{c21} & D_{21}^T \end{bmatrix}, \\
R_\times &= \begin{bmatrix} V_{c21}\left(V_{c11}^TV_{c11}\right)^{-1}V_{c21}^T - I & V_{c21}\left(V_{c11}^TV_{c11}\right)^{-1}D_{21}^T \\ D_{21}\left(V_{c11}^TV_{c11}\right)^{-1}V_{c21}^T & D_{21}\left(V_{c11}^TV_{c11}\right)^{-1}D_{21}^T \end{bmatrix}.
\end{aligned} \tag{3.75}$$

Also let

$$J_\times = \begin{bmatrix} -I_{m_2} & 0 \\ 0 & I_{p_2} \end{bmatrix}. \tag{3.76}$$

Now we shall state the second set of necessary solvability conditions by the aid of the next theorem.

Theorem 3.5 *If the H^∞-control problem formulated for T has a solution then the KPYS($\Sigma_\times, J_\times$) has a stabilizing solution*

$$(Z, V_\times, W_\times) = \left(Z, \begin{bmatrix} V_{\times 11} & 0 \\ V_{\times 21} & V_{\times 22} \end{bmatrix}, \begin{bmatrix} W_{\times 1} \\ W_{\times 2} \end{bmatrix}\right)$$

where $Z \geq 0$ and the partitions of $V_\times$ and $W_\times$ are taken in accordance with $J_\times$ in (3.76).

Proof. Notice first that the regularity conditions (R2) hold for T_o. Indeed $D_{21}V_{c11}^{-1}$ is of full row rank and

$$\begin{aligned} & \operatorname{rank} \begin{bmatrix} j\omega I - A - B_1F_1 & -B_1V_{c11}^{-1} \\ C_2 + D_{21}F_1 & D_{21}V_{c11}^{-1} \end{bmatrix} \\ = \ & \operatorname{rank} \begin{bmatrix} j\omega I - A & -B_1 \\ C_2 & D_{21} \end{bmatrix} = n + p_2 \quad \forall\, \omega \in \mathbf{R}. \end{aligned}$$

Hence by invoking Proposition 3.3 in conjunction with Theorem 3.3, in which Σ_o has been replaced by $\Sigma_\times$, the conclusion follows. □

Assume now that the normalizing conditions (3.52) hold. Then $V_c = I_m$, $W_c^T = X \begin{bmatrix} B_1 & B_2 \end{bmatrix}$ and $F^T = \begin{bmatrix} XB_1 & -XB_2 \end{bmatrix}$, as easily can be checked. In this case (3.75) yields

$$\begin{aligned} A_\times &= A^T + XB_1B_1^T, \\ B_\times &= \begin{bmatrix} XB_2 & C_2^T \end{bmatrix}, \\ Q_\times &= B_1B_1^T, \\ L_\times &= \begin{bmatrix} 0 & 0 \end{bmatrix}, \\ R_\times &= \begin{bmatrix} -I_{m_2} & 0 \\ 0 & I_{p_2} \end{bmatrix} = J_\times. \end{aligned} \tag{3.77}$$

Since $R_\times = J_\times$ in (3.77), it follows that the KPYS($\Sigma_\times, J_\times$) is equivalent to the ARE($\Sigma_\times$) which is

$$\begin{aligned} & \left(A + B_1B_1^TX\right) Z + Z \left(A + B_1B_1^TX\right)^T \\ & \quad -Z\left(-XB_2B_2^TX + C_2^TC_2\right) Z + B_1B_1^T = 0. \end{aligned} \tag{3.78}$$

Thus we have:

Theorem 3.6 *If the H^∞-control problem formulated for T has a solution then the ARE (3.78) has a positive semidefinite stabilizing solution Z.* □

By combining Theorems 3.2 and 3.5 we obtain the main necessity result on the solvability of the H^∞-control problem.

Theorem 3.7 *If the H^∞-control problem formulated for T in (3.1) has a solution (3.2), then both KPYS(Σ_c, J_c) and KPYS($\Sigma_\times, J_\times$) have stabilizing solutions*

$$(X, V_c, W_c) = \left(X, \begin{bmatrix} V_{c11} & 0 \\ V_{c21} & V_{c22} \end{bmatrix}, \begin{bmatrix} W_{c1} \\ W_{c2} \end{bmatrix}\right)$$

and

$$(Z, V_\times, W_\times) = \left(Z, \begin{bmatrix} V_{\times 11} & 0 \\ V_{\times 21} & V_{\times 22} \end{bmatrix}, \begin{bmatrix} W_{\times 1} \\ W_{\times 2} \end{bmatrix}\right),$$

respectively with $X, Z \geq 0$. Here Σ_c, J_c and $\Sigma_\times$, $J_\times$ are given by (3.23), (3.39) and (3.75), (3.76), respectively and the partitions of V_c, W_c and $V_\times$, $W_\times$ are taken in accordance with J_c and $J_\times$ in (3.39) and (3.76), respectively. □

Similarly we have:

Theorem 3.8 *Assume that the normalizing conditions (3.52) hold. In this situation, if the H^∞-control problem formulated for T in (3.1) has a solution (3.2) then both ARE(Σ_c) and ARE($\Sigma_\times$), made explicit by (3.55) and (3.78), have positive semi-definite stabilizing solutions X and Z, respectively.* □

Remark 3.1 Since the input data of the KPYS($\Sigma_\times, J_\times$) are delivered by the stabilizing solution to the KPYS(Σ_c, J_c), the necessary solvability conditions stated in Theorem 3.7 will be termed as *coupled* ones. At the same time, Theorems 3.2 and 3.3 state a set of necessity conditions which are apparently *uncoupled*. The connections between these two sets of necessity conditions will be investigated in the next section. □

3.5. COUPLED AND UNCOUPLED NECESSITY SOLVABILITY CONDITIONS

In this section the development is essentially based on the theory developed in Section 2.4. The main result is the following:

Theorem 3.9 *Assume that the KPYS(Σ_c, J_c) has a stabilizing solution*

$$(X, V_c, W_c) = \left(X, \begin{bmatrix} V_{c11} & 0 \\ V_{c21} & V_{c22} \end{bmatrix}, \begin{bmatrix} W_{c1} \\ W_{c2} \end{bmatrix}\right)$$

with $X \geq 0$ and V_c and W_c partitioned in accordance with J_c in (3.39). Then the following two statements are equivalent:

(1) The KPYS($\Sigma_\times, J_\times$) has a stabilizing solution

$$(Z, V_\times, W_\times) = \left(Z, \begin{bmatrix} V_{\times 11} & 0 \\ V_{\times 21} & V_{\times 22} \end{bmatrix}, \begin{bmatrix} W_{\times 1} \\ W_{\times 2} \end{bmatrix}\right)$$

with $Z \geq 0$ and $V_\times$ and $W_\times$ partitioned in accordance with $J_\times$ in (3.76).

(2) The KPYS(Σ_o, J_o) has a stabilizing solution

$$(Y, V_o, W_o) = \left(Y, \begin{bmatrix} V_{o11} & 0 \\ V_{o21} & V_{o22} \end{bmatrix}, \begin{bmatrix} W_{o1} \\ W_{o2} \end{bmatrix}\right)$$

with $Y \geq 0$ and V_o and W_o partitioned in accordance with J_o in (3.51). In addition

$$\rho(XY) < 1. \tag{3.79}$$

If (1) holds then

$$Z = Y(I - XY)^{-1}. \tag{3.80}$$

Proof. As $R_o = V_o^T J_o V_o$, $R_\times = V_\times^T J_\times V_\times$, $R_{o,22} > 0$, $R_{\times,22} > 0$ and V_o and $V_\times$ are both of lower left block triangular form it is easy to notice that the statements (1) and (2) are equivalent to the following two pairs of assertions:

(1a) The ARE($\Sigma_\times$) has a positive semidefinite stabilizing solution Z.

(1b)

$$\operatorname{sgn} R_\times = J_\times$$

and

(2a) The ARE(Σ_o) has a positive semidefinite stabilizing solution Y and (3.79) holds.

(2b)

$$\operatorname{sgn} R_o = J_o.$$

Let $\lambda M_{\Sigma_\times} - N_{\Sigma_\times}$ and $\lambda M_{\Sigma_o} - N_{\Sigma_o}$ be the EHP($\Sigma_\times$) and EHP(Σ_o) given explicitly by

$$M_{\Sigma_\times} = \begin{bmatrix} I_n & & & \\ & I_n & & \\ & & 0_{m_2} & \\ & & & 0_{p_2} \end{bmatrix}, \tag{3.81}$$

$$N_{\Sigma_\times} = \begin{bmatrix} N_{11}^\times & N_{12}^\times \\ N_{21}^\times & N_{22}^\times \end{bmatrix}, \tag{3.82}$$

with

$$
\begin{aligned}
N_{11}^{\times} &:= \begin{bmatrix} A^T - W_{c1}^T V_{c11}^{-T} B_1^T & 0 \\ -B_1 V_{c11}^{-1} V_{c11}^{-T} B_1^T & -A + B_1 V_{c11}^{-1} W_{c1} \end{bmatrix}, \\
N_{12}^{\times} &:= \begin{bmatrix} -W_{c1}^T V_{c11}^{-T} V_{c21}^T + W_{c2}^T & C_2^T - W_{c1}^T V_{c11}^{-T} D_{21}^T \\ -B_1 V_{c11}^{-1} V_{c11}^{-T} V_{c21}^T & -B_1 V_{c11}^{-1} V_{c11}^{-T} D_{21}^T \end{bmatrix}, \\
N_{21}^{\times} &:= \begin{bmatrix} V_{c21} V_{c11}^{-1} V_{c11}^{-T} B_1^T & -V_{c21} V_{c11}^{-1} W_{c1} + W_{c2} \\ D_{21} V_{c11}^{-1} V_{c11}^{-T} B_1^T & C_2 - D_{21} V_{c11}^{-1} W_{c1} \end{bmatrix}, \\
N_{22}^{\times} &:= \begin{bmatrix} V_{c21} V_{c11}^{-1} V_{c11}^T V_{c21}^T - I_{m_2} & V_{c21} V_{c11}^{-1} V_{c11}^{-T} D_{21}^T \\ D_{21} V_{c11}^{-1} V_{c11}^{-T} V_{c21}^T & D_{21} V_{c11}^{-1} V_{c11}^{-T} D_{21}^T \end{bmatrix},
\end{aligned}
$$

and

$$
M_{\Sigma_o} = \begin{bmatrix} I_n & & & \\ & I_n & & \\ & & 0_{p_1} & \\ & & & 0_{p_2} \end{bmatrix}, \tag{3.83}
$$

$$
N_{\Sigma_o} = \begin{bmatrix} N_{11}^o & N_{12}^o \\ N_{21}^o & N_{22}^o \end{bmatrix}, \tag{3.84}
$$

with

$$
\begin{aligned}
N_{11}^o &:= \begin{bmatrix} A^T & 0 \\ -B_1 B_1^T & -A \end{bmatrix}, \\
N_{12}^o &:= \begin{bmatrix} C_1^T & C_2^T \\ -B_1 D_{11}^T & -B_1 D_{21}^T \end{bmatrix}, \\
N_{21}^o &:= \begin{bmatrix} D_{11} B_1^T & C_1 \\ D_{21} B_1^T & C_2 \end{bmatrix}, \\
N_{22}^o &:= \begin{bmatrix} D_{11} D_{11}^T - I_{p_1} & D_{11} D_{21}^T \\ D_{21} D_{11}^T & D_{21} D_{21}^T \end{bmatrix},
\end{aligned}
$$

as directly follows from $\Sigma_\times$ and Σ_o in (3.75) and (3.50), respectively.

Let the following augmented versions $\lambda M_a^\times - N_a^\times$ and $\lambda M_a^o - N_a^o$ of $\lambda M_{\Sigma_\times} - N_{\Sigma_\times}$ and $\lambda M_{\Sigma_o} - N_{\Sigma_o}$ be introduced by

$$
M_a^\times = \begin{bmatrix} M_{\Sigma_\times} & & \\ & 0_{m_1} & \\ & & 0_{p_1} \end{bmatrix}, N_a^\times = \begin{bmatrix} N_{\Sigma_\times} & & \\ & -V_{c11}^{-1} V_{c11}^{-T} & \\ & & -I_{p_1} \end{bmatrix}, \tag{3.85}
$$

and

$$M_a^o = \begin{bmatrix} M_{\Sigma_o} & & \\ & 0_{m_1} & \\ & & 0_{m_2} \end{bmatrix}, \quad N_a^o = \begin{bmatrix} N_{\Sigma_o} & & \\ & -I_{m_1} & \\ & & -I_{m_2} \end{bmatrix}, \tag{3.86}$$

respectively. Notice that $M_a^\times$, $N_a^\times$, M_a^o and N_a^o are all of the same dimension $2n+m+p$. Let also the following $(2n+m+p)\times(2n+m+p)$ nonsingular matrices

$$U_a = \begin{bmatrix} U_{a11} & U_{a12} \\ 0 & U_{a22} \end{bmatrix}, \tag{3.87}$$

with

$$\begin{aligned} U_{a11} &:= \begin{bmatrix} I_n & X \\ 0 & I_n \end{bmatrix}, \\ U_{a12} &:= \begin{bmatrix} W_{c2}^T & 0 & C_1^T D_{11} & -C_1^T \\ -B_1 V_{c21}^T & 0 & -B_1 + B_1\Delta & 0 \end{bmatrix}, \\ U_{a22} &:= \begin{bmatrix} D_{11}V_{c21}^T & 0 & -D_{11}\Delta & I_{p_1} - D_{11}D_{11}^T \\ D_{21}V_{c21}^T & I_{p_2} & D_{21} - D_{21}\Delta & -D_{21}^T D_{11} \\ -V_{c21}^T & 0 & \Delta & D_{11}^T \\ I_{m_2} & 0 & V_{c21} & 0 \end{bmatrix}, \end{aligned}$$

and

$$W_a = \begin{bmatrix} W_{a11} & 0 \\ W_{a21} & W_{a22} \end{bmatrix}, \tag{3.88}$$

with

$$\begin{aligned} W_{a11} &:= \begin{bmatrix} I_n & -X \\ 0 & I_n \end{bmatrix}, \\ W_{a21} &:= \begin{bmatrix} V_{c21}B_1^T & W_{c2} \\ 0 & 0 \\ B_1^T - \Delta B_1^T & D_{11}^T C_1 \\ -D_{11}B_1^T & -C_1 \end{bmatrix}, \\ W_{a22} &:= \begin{bmatrix} V_{c21}D_{11}^T & V_{c21}D_{21}^T & -V_{c21} & I_{m_2} \\ 0 & I_{p_2} & 0 & 0 \\ -\Delta D_{11}^T & -\Delta D_{21}^T + D_{21}^T & \Delta & V_{c21}^T \\ I_{p_1} - D_{11}D_{11}^T & -D_{11}D_{21}^T & D_{11} & 0 \end{bmatrix}, \end{aligned}$$

be introduced. Here

$$\Delta := V_{c11}^T V_{c11} - V_{c21}^T V_{c21} = \Delta^T.$$

Using the KPYS(Σ_c, J_c) one can check after simple computations that

$$U_a M_a^\times W_a = M_a^o \tag{3.89}$$

and

$$U_a N_a^\times W_a = N_a^o, \tag{3.90}$$

i.e., $\lambda M_a^\times - N_a^\times \sim \lambda M_a^o - N_a^o$.

Now we shall prove the implication (1a)$\Rightarrow$(1b). Since the ARE$(\Sigma_\times)$ has a stabilizing solution we can write (see Theorem 2.7)

$$N_{\Sigma_\times} V^\times = M_{\Sigma_\times} V^\times S^\times, \tag{3.91}$$

where

$$V^\times = \begin{bmatrix} I \\ Z \\ F^\times \end{bmatrix}. \tag{3.92}$$

In (3.91), (3.92) $Z \geq 0$ and $F^\times$ are the stabilizing solution to the ARE$(\Sigma_\times)$ and the associated stabilizing feedback gain, respectively, and $S^\times \in \mathbf{R}^{n\times n}$ is stable. Let

$$V_a^\times := \begin{bmatrix} V^\times \\ 0 \\ 0 \end{bmatrix} \begin{matrix} \\ \}m_1 \\ \}p_1 \end{matrix} \tag{3.93}$$

With (3.85) and (3.93) the augmented version of (3.91) becomes

$$N_a^\times V_a^\times = M_a^\times V_a^\times S^\times. \tag{3.94}$$

Using now (3.89) and (3.90), (3.94) yields

$$N_a^o W_a^{-1} V_a^\times = M_a^o W_a^{-1} V_a^\times S^\times. \tag{3.95}$$

In accordance with the structure of W_a in (3.88) one obtains

$$W_a^{-1} = \begin{bmatrix} W_{a11}^{-1} & 0 \\ * & W_{a22}^{-1} \end{bmatrix}, \tag{3.96}$$

where

$$W_{a11}^{-1} = \begin{bmatrix} I_n & X \\ 0 & I_n \end{bmatrix}. \tag{3.97}$$

Using (3.96), (3.97) one obtains that

$$W_a^{-1} V_a^\times = \begin{bmatrix} I + XZ \\ Z \\ * \end{bmatrix} \begin{matrix} \\ \\ \}m+p \end{matrix} \tag{3.98}$$

Since both X and Z are positive semidefinite, $I+XZ$ is nonsingular. Using this conclusion and substituting (3.98) in (3.95) one obtains

$$N_a^o \begin{bmatrix} I \\ Y \\ * \end{bmatrix}_{\}m+p} = M_a^o \begin{bmatrix} I \\ Y \\ * \end{bmatrix}_{\}m+p} S^o, \tag{3.99}$$

where

$$Y := Z(I+XZ)^{-1} \geq 0 \tag{3.100}$$

and

$$S^o := (I+XZ)S^\times (I+XZ)^{-1}, \tag{3.101}$$

which is clearly stable. Now it is easy to see that taking into account (3.86), (3.99) yields

$$N_{\Sigma_o} \begin{bmatrix} I \\ Y \\ * \end{bmatrix}_{\}p} = M_{\Sigma_o} \begin{bmatrix} I \\ Y \\ * \end{bmatrix}_{\}p} S^o. \tag{3.102}$$

Invoking again Theorem 2.7, (3.102) with (3.101) show that Y defined by (3.100) is exactly the positive semidefinite stabilizing solution to the ARE(Σ_o). Furthermore (3.100) yields

$$I - XY = (I+XZ)^{-1} > 0 \tag{3.103}$$

from which (3.79) follows. Using (3.108), (3.80) follows from (3.100) automatically.

The proof of the implication (1b)⇒(1a) follows by reversing the above arguments.

Let us prove now that (1b)⇔(2b) To this end notice that

$$W_{a22} = U_{a22}^T,$$

as follows from (3.87) and (3.88). Then (3.90) yields

$$U_{a22} \begin{bmatrix} N_{22}^\times & & \\ & -V_{c11}^{-1}V_{c11}^{-T} & \\ & & -I_{p_1} \end{bmatrix} U_{a22}^T$$
$$= \begin{bmatrix} N_{22}^o & & \\ & -I_{m_1} & \\ & & -I_{m_2} \end{bmatrix}. \tag{3.104}$$

Taking into account the preservation of the signature, (3.104) implies that

$$\operatorname{sgn} N_{22}^\times = \begin{bmatrix} -I_{m_2} & \\ & I_{p_2} \end{bmatrix} \Leftrightarrow \operatorname{sgn} N_{22}^o = \begin{bmatrix} -I_{p_1} & \\ & I_{p_2} \end{bmatrix}.$$

Since $R_\times = N_{22}^\times$ and $R_o = N_{22}^o$ the conclusion follows. Thus the proof is completed. □

3.6. SUFFICIENT SOLVABILITY CONDITIONS: AN EXPLICIT SOLUTION IN KPYS TERMS

In this section we shall show that the necessity conditions stated in Theorem 3.7 are also sufficient conditions for solving the H^∞-control problem. Thus the main result of this section is stated as follows.

Theorem 3.10 *Let the system (3.1) be given and assume that the following conditions all hold:*

(1) For Σ_c and J_c specified in (3.29) and (3.39), the KPYS(Σ_c, J_c) has a stabilizing solution

$$(X, V_c, W_c) = \left(X, \begin{bmatrix} V_{c11} & 0 \\ V_{c21} & V_{c22} \end{bmatrix}, \begin{bmatrix} W_{c1} \\ W_{c2} \end{bmatrix} \right),$$

where $X \geq 0$ and the partitions of V_c and W_c are taken in accordance with J_c in (3.39).

(2) For $\Sigma_\times$ and $J_\times$ given explicitly in (3.75) and (3.76), the KPYS($\Sigma_\times, J_\times$) has a stabilizing solution

$$(Z, V_\times, W_\times) = \left(Z, \begin{bmatrix} V_{\times 11} & 0 \\ V_{\times 21} & V_{\times 22} \end{bmatrix}, \begin{bmatrix} W_{\times 1} \\ W_{\times 2} \end{bmatrix} \right),$$

where $Z \geq 0$ and $V_\times$ and $W_\times$ are partitioned in accordance with $J_\times$ in (3.76).

Then there exists a solution K (see (3.2)) of the H^∞-control problem. Furthermore the data of K (for $\gamma = 1$) are given by the following explicit formulae:

$$\begin{aligned} A_K &= A + B_1 F_1 + B_2 F_2 + B_K C_{2F_1}, \\ B_K &= -\left(B_1 S D_{21}^T + Z C_{2F_1}^T\right)\left(D_{21} S D_{21}^T\right)^{-1} - B_2 D_K, \\ C_K &= -F_2 + D_K C_{2F_1}, \\ D_K &= -\left(D_{12}^T D_{12}\right)^{-1} D_{12}^T D_{11} S D_{21}^T \left(D_{21} S D_{21}^T\right)^{-1}, \end{aligned} \tag{3.105}$$

where

$$\begin{aligned} \begin{bmatrix} F_1 \\ F_2 \end{bmatrix} &= \begin{bmatrix} -V_{c11}^{-1} W_{c1} \\ V_{c22}^{-1} V_{c21} V_{c11}^{-1} W_{c1} - V_{c22}^{-1} W_{c2} \end{bmatrix}, \\ C_{2F_1} &= C_2 + D_{21} F_1, \\ S &= \left(V_{c11}^T V_{c11}\right)^{-1}. \end{aligned} \tag{3.106}$$

□

Notice that (3.105) are given in terms of scaled data. In order to obtain the original ones use (3.16), (3.17).

Remark 3.2 If the normalizing conditions (3.52) hold then the formulae (3.105) become much simpler. Indeed, (3.106) gives

$$\begin{aligned}\begin{bmatrix} F_1 \\ F_2 \end{bmatrix} &= \begin{bmatrix} B_1^T X \\ -B_2^T X \end{bmatrix}, \\ C_{2F_1} &= C_2, \\ S &= I,\end{aligned}$$

from which (3.105) become

$$\begin{aligned} A_K &= A + \left(B_1 B_1^T - B_2 B_2^T\right) X - Z C_2^T C_2, \\ B_K &= -Z C_2^T, \\ C_K &= B_2^T X, \\ D_K &= 0. \end{aligned} \tag{3.107}$$

□

In order to prove Theorem 3.10, in fact to obtain an explicit solution of the H^∞-control problem, we shall solve it successively for some simpler cases, and in the end we reduce the general case to these particular ones. Thus we shall start with the simplest case.

1. The Disturbance Estimation Problem (DEP)
We assume that:

(DE1) D_{12} is square and nonsingular;

(DE2) $A - B_2 D_{12}^{-1} C_1$ is stable;

(DE3) D_{21} is square and nonsingular;

(DE4) $A - B_1 D_{21}^{-1} C_2$ is stable.

Theorem 3.11 *A solution K to the DEP is given by:*

$$\begin{aligned} A_K &= A - B_1 D_{21}^{-1} C_2 - B_2 D_{12}^{-1} C_1 + B_2 D_{12}^{-1} D_{11} D_{21}^{-1} C_2, \\ B_K &= \left(B_1 - B_2 D_{12}^{-1} D_{11}\right) D_{21}^{-1}, \\ C_K &= -D_{12}^{-1} \left(C_1 - D_{11} D_{21}^{-1} C_2\right), \\ D_K &= -D_{12}^{-1} D_{11} D_{21}^{-1}. \end{aligned} \tag{3.108}$$

Proof. Assume first that

$$D_{11} = 0 \tag{3.109}$$

and let $F_2 := -D_{12}^{-1}C_1$, $K_2 = -B_1D_{21}^{-1}$. Then $A + B_2F_2$ and $A + KC_2$ are stable and we can construct the Kalman compensator (defined in Section 1.2)

$$\begin{aligned} K &= \left[\begin{array}{c|c} A + B_2F_2 + K_2C_2 & -K_2 \\ \hline F_2 & 0 \end{array}\right] \\ &= \left[\begin{array}{c|c} A - B_2D_{12}^{-1}C_1 - B_1D_{21}^{-1}C_2 & B_1D_{21}^{-1} \\ \hline -D_{12}^{-1}C_1 & 0 \end{array}\right], \qquad (3.110) \\ u_2 &= Ky_2. \end{aligned}$$

Using (3.3), (3.4) one obtains

$$\begin{aligned} T_{y_1u_1} &= \left[\begin{array}{cc|c} A & -B_2D_{12}^{-1}C_1 & B_1 \\ B_1D_{21}^{-1}C_2 & A - B_2D_{12}^{-1}C_1 - B_1D_{21}^{-1}C_2 & B_2 \\ \hline C_1 & -C_1 & 0 \end{array}\right] \\ &= \left[\begin{array}{cc|c} A - B_2D_{12}^{-1}C_1 & -B_2D_{12}^{-1}C_1 & B_1 \\ 0 & A - B_1D_{21}^{-1}C_2 & 0 \\ \hline 0 & -C_1 & 0 \end{array}\right] \\ &= 0, \qquad (3.111) \end{aligned}$$

where the state space coordinate change

$$\left[\begin{array}{c} \hat{x} \\ \hat{x}_K \end{array}\right] = \left[\begin{array}{cc} I & 0 \\ -I & I \end{array}\right] \left[\begin{array}{c} x \\ x_K \end{array}\right]$$

has been performed. By inspecting now the rightmost term in (3.111) the conclusion follows. Assume now that (3.109) failed. Then perform the pre-output feedback low

$$\begin{aligned} u_2 &= -D_{12}^{-1}D_{11}D_{21}^{-1}y_2 + \tilde{u}_2 \\ &= -D_{12}^{-1}D_{11}D_{21}^{-1}C_2x - D_{12}^{-1}D_{11}u_1 + \tilde{u}_2, \qquad (3.112) \end{aligned}$$

and (3.1) is changed into

$$\begin{aligned} \widetilde{T} &= \left[\begin{array}{c|cc} A - B_2D_{12}^{-1}D_{11}D_{21}^{-1}C_2 & B_1 - B_2D_{12}^{-1}D_{11} & B_2 \\ \hline C_1 - D_{11}D_{21}^{-1}C_2 & 0 & D_{12} \\ C_2 & D_{21} & 0 \end{array}\right] \\ &= \left[\begin{array}{c|cc} \widetilde{A} & \widetilde{B}_1 - B_2D_{12}^{-1}D_{11} & B_2 \\ \hline \widetilde{C}_1 & 0 & D_{12} \\ C_2 & D_{21} & 0 \end{array}\right], \qquad (3.113) \\ \left[\begin{array}{c} y_1 \\ y_2 \end{array}\right] &= \widetilde{T}\left[\begin{array}{c} u_1 \\ \tilde{u}_2 \end{array}\right]. \end{aligned}$$

It is easily checked that conditions (DE1)–(DE4) hold with respect to $\widetilde{T}$ and, in addition, (3.109) holds as well. Using (3.110) one obtains

$$\begin{aligned} \widetilde{K} &= \left[\begin{array}{c|c} \widetilde{A} - B_2 D_{12}^{-1}\widetilde{C}_1 - \widetilde{B}_1 D_{21}^{-1} C_2 & \widetilde{B}_1 D_{21}^{-1} \\ \hline -D_{12}^{-1}\widetilde{C}_1 & 0 \end{array}\right], \qquad (3.114) \\ \tilde{u}_2 &= \widetilde{K} y_2. \end{aligned}$$

By combining (3.112) with (3.114), the formulae (3.108) follow trivially. □

2. The Disturbance Feedforward Problem (DFP)
We assume that:

(DF1) D_{21} is square and nonsingular;

(DF2) $A - B_1 D_{21}^{-1} C_2$ is stable;

(DF3) Condition (1) in Theorem 3.10 holds.

Theorem 3.12 *A solution K to the DFP is given by*

$$\begin{aligned} A_K &= A - B_1 D_{21}^{-1} C_2 \\ &\quad - B_2 \left(D_{12}^T D_{12}\right)^{-1} \left[D_{12}^T \left(C_1 - D_{11} D_{21}^{-1} C_2\right) + B_2^T X\right], \\ B_K &= \left(B_1 - B_2 \left(D_{12}^T D_{12}\right)^{-1} D_{12}^T D_{11}\right) D_{21}^{-1}, \qquad (3.115) \\ C_K &= -\left(D_{12}^T D_{12}\right)^{-1} \left[D_{12}^T \left(C_1 - D_{11} D_{21}^{-1} C_2\right) + B_2^T X\right], \\ D_K &= -\left(D_{12}^T D_{12}\right)^{-1} D_{12}^T D_{11} D_{21}^{-1}. \end{aligned}$$

Proof. Assumption DF3 allows us to consider the system T_o given in (3.67), that is

$$\begin{aligned} T_o &= \left[\begin{array}{c|cc} A + B_1 F_1 & B_1 V_{c11}^{-1} & B_2 \\ \hline -F_{c22} F_2 & V_{c21} V_{c11}^{-1} & V_{c22} \\ C_2 + D_{21} F_1 & D_{21} V_{c11}^{-1} & 0 \end{array}\right] \\ &= \left[\begin{array}{c|cc} A_o & B_{o1} & B_{o2} \\ \hline C_{o1} & D_{o11} & D_{o12} \\ C_{o2} & D_{o21} & D_{o22} \end{array}\right]. \qquad (3.116) \end{aligned}$$

Invoking now Proposition 3.3 the desired solution coincides with the solution to the H^∞-control problem formulated for T_o. Notice that (DE1)–(DE4) hold with respect to T_o. Indeed, both D_{o12} and D_{o21} are nonsingular, and

$$\begin{aligned} A_o - B_{o1} D_{o21}^{-1} C_{02} &= A + B_1 F_1 - B_1 D_{21}^{-1} (C_2 + D_{21} F_1) \\ &= A - B_1 D_{21}^{-1} C_2 \end{aligned}$$

and

$$A_o - B_{o2}D_{o12}^{-1}C_{o1} = A + B_1F_1 + B_2F_2 = A + BF$$

are stable. Thus by updating (3.108) with data given by (3.116), the formulae (3.115) are easily recovered. □

3. The Output Estimation Problem (OEP)

We assume that:

(OE1) D_{12} is square and nonsingular;

(OE2) $A - B_2D_{12}^{-1}C_1$ is stable;

(OE3) The dual of condition (1) in Theorem 3.10 holds, that is the KPYS(Σ_o, J_o), with Σ_o and J_o given in (3.50) and (3.51), has a stabilizing solution

$$(Y, V_o, W_o) = \left(Y, \begin{bmatrix} V_{o11} & 0 \\ V_{o21} & V_{o22} \end{bmatrix}, \begin{bmatrix} W_{o1} \\ W_{o2} \end{bmatrix}\right),$$

where $Y \geq 0$ and V_o and W_o are partitioned in accordance with J_o in (3.51).

Theorem 3.13 *A solution K of the OEP is given by*

$$\begin{aligned} A_K &= A - B_2D_{12}^{-1}C_1 \\ &\quad - \left[\left(B_1 - B_2D_{12}^{-1}D_{11}\right)D_{21}^T + YC_2^T\right]\left(D_{21}D_{21}^T\right)^{-1}C_2, \\ B_K &= -\left[\left(B_1 - B_2D_{12}^{-1}D_{11}\right)D_{21}^T + YC_2^T\right]\left(D_{21}D_{21}^T\right)^{-1}, \\ C_K &= D_{12}^{-1}\left(C_1 - D_{11}D_{21}^T\left(D_{21}D_{21}^T\right)^{-1}C_2\right), \\ D_K &= -D_{12}^{-1}D_{11}D_{21}^T\left(D_{21}D_{21}^T\right)^{-1}. \end{aligned} \tag{3.117}$$

□

This theorem is the dual of Theorem 3.12.

4. The general case: The proof of Theorem 3.10

Using condition (1) in Theorem 3.10 we can write the system T_o (see (3.67))

$$\begin{aligned} T_o &= \left[\begin{array}{c|cc} A + B_1F_1 & B_1V_{c11}^{-1} & B_2 \\ \hline -F_{c22}F_2 & V_{c21}V_{c11}^{-1} & V_{c22} \\ C_2 + D_{21}F_1 & D_{21}V_{c11}^{-1} & 0 \end{array}\right] \\ &= \left[\begin{array}{c|cc} A_o & B_{o1} & B_{o2} \\ \hline C_{o1} & D_{o11} & D_{o12} \\ C_{o2} & D_{o21} & D_{o22} \end{array}\right], \end{aligned} \tag{3.118}$$

which satisfies all the assumptions of the OEP. Indeed, (OE1) and (OE2) are obvious as they are the same as in the proof of Theorem 3.12. Condition

(OE3) refers to the KPYS(Σ_o, J_o) where Σ_o, J_o are updated with the data of (3.118). But this is exactly the KPYS($\Sigma_\times, J_\times$) which has a stabilizing solution as it is mentioned in condition (2) of Theorem 3.10. Hence using the formulae (3.117) updated with the data of (3.118) we obtain the formulae (3.105). Invoking now Proposition 3.3 we conclude that (3.118) are exactly the data of the desired controller. Thus the proof is completed. □

3.7. SINGULAR PROBLEMS: NECESSARY AND SUFFICIENT CONDITIONS

The developments performed in the previuos sections essentially use the regularity assumptions (R1) and (R2) adopted in Section 3.1. If these conditions are fulfilled it is said that the H^∞-control problem is *regular*. In the case when at least one of these assumptions do not hold, the H^∞-control problem is called *singular* and specific solvability methods are required. One of these methods which received much attention in the recent years is based on *linear matrix inequalities* (LMIs). Some details are given in Notes and References. It is not our intention to detail here the LMIs-based methodology but in the following we shall state two known results which play a central role in this approach.

Consider the H^∞ problem stated in Section 3.1 for which the regularity conditions (R1) and (R2) are no more necessarily assumed. Then the following result holds

Theorem 3.14 *The H^∞-control problem has a solution if and only if there exist the positive definite symmetric matrices R and S such that*

$$\begin{bmatrix} \mathcal{N}_{12} & 0 \\ 0 & I \end{bmatrix}^T \begin{bmatrix} AR+RA^T & RC_1^T & \vdots & B_1 \\ C_1R & -\gamma I & \vdots & D_{11} \\ \cdots & \cdots & \cdots & \cdots \\ B_1^T & D_{11}^T & \vdots & -\gamma I \end{bmatrix} \begin{bmatrix} \mathcal{N}_{12} & 0 \\ 0 & I \end{bmatrix} < 0, \quad (3.119)$$

$$\begin{bmatrix} \mathcal{N}_{21} & 0 \\ 0 & I \end{bmatrix}^T \begin{bmatrix} A^TS+SA & SB_1 & \vdots & C_1^T \\ B_1^TS & -\gamma I & \vdots & D_{11}^T \\ \cdots & \cdots & \cdots & \cdots \\ C_1 & D_{11} & \vdots & -\gamma I \end{bmatrix} \begin{bmatrix} \mathcal{N}_{21} & 0 \\ 0 & I \end{bmatrix} < 0, \quad (3.120)$$

$$\lambda_{\min}(RS) > 1, \quad (3.121)$$

where $\mathcal{N}_{12}$ and $\mathcal{N}_{21}$ denote bases of the null spaces of $\begin{bmatrix} B_2^T & D_{12}^T \end{bmatrix}$ and $\begin{bmatrix} C_2 & D_{21} \end{bmatrix}$ respectively, and $\lambda_{\min}(\cdot)$ is the smallest eigenvalue of $(\cdot)$. □

The following proposition is a technical result which allows to replace the LMIs (3.119) and (3.220) by Riccati inequations explicitly depending on the generalized system data.

Proposition 3.4 (Finsler's Lemma) *Let Ω be an $n \times n$ symmetric matrix, P an $m \times n$ matrix and $\mathcal{N}$ a basis of the null space of P. Then $\mathcal{N}^T \Omega \mathcal{N} < 0$ if and only if there exists a constant $\alpha > 0$ such that $\Omega - \alpha P^T P < 0$.* □

The conditions stated in Theorem 3.14 are tested by appropriate numerical algorithms to solve feasibility and optimization problems. If these conditions are fulfilled then an H^∞ controller is determined in terms of the solutions to the LMIs. This controller is obtained as a solution of an LMI with respect to the controller parameters or by using explicit formulae.

Based on the two results stated above we shall derive in Chapter 6 specific solvability conditions and explicit formulae of the controller for some particular singular H^∞ problems frequently arising in control applications.

NOTES AND REFERENCES

The H^∞-control problem originated in (Zames, 1981) has received much attention in the last two decades. For an extensive historical perspective of the problem, see for instance (Doyle *et al.*, 1989). Different methods have been proposed for solving this problem. We mention among them those based on interpolation and operator theory, intensively used particularly in the early period (Adamian *et al.*, 1978), (Ball and Helton, 1983), reduction to the Nehari problem (Francis, 1986), polynomial approaches (Kwakernaak, 1986), (Grimble, 1988) and state space methods. A complete state space solution of the regular H^∞-control problem in the continuous time case is derived in (Glover and Doyle, 1988) and (Doyle *et al.*, 1989) in terms of two AREs; explicit formulae for the parameterization of all solutions are also given. A discrete time counterpart of the two-Riccati method is presented in (Limebeer *et al.*, 1989), (Stoorvogel, 1990), (Iglesias, 1991) and (Halanay and Ionescu, 1994). The state space solution using the generalized Popov–Yakubovich theory is obtained in (Ionescu and Weiss, 1993) whose main ideas are followed by the developments performed in this chapter for the continuous time case.

For the singular H^∞-control problems, we cite as early references (Petersen, 1987), (Khargonekar *et al.*, 1988), (Zhou and Khargonekar, 1988), (Stoorvogel and Trentelman, 1990) in which the state feedback case is considered. Further extensions to the output feedback case have been derived in (Sampei *et al.*, 1990) and (Scherer, 1992) where necessary and sufficient solvability conditions are given in terms of two Riccati inequalities depending on the controller parameters. The proof of these results also include explicit formulae for the controller. Independent necessary and sufficient conditions with respect to the controller parameters for the general H^∞-control problem has been expressed in terms of LMIs. A comprehensive treatment of this method can be found for instance in (Gahinet and Apkarian, 1994), (Boyd *et al.*, 1994), (Iwasaki and Skelton, 1994). At the same time, powerful numerical algorithms and appropriate software packages for solving LMI optimization problems have been developed (*e.g.*, (Gahinet *et al.*, 1995)).

Explicit formulae for the parameterization of all solutions to the general H^∞-control problem have been derived in (Iwasaki and Skelton, 1994).

CHAPTER 4

THE NEHARI PROBLEM

Over the last years, the Nehari problem has been receiving considerable attention. This interest is because many control problems including H^∞ optimization, robustness with respect to dynamic modeling uncertainty, disturbance attenuation, and mixed sensitivity may be reduced to Nehari problems.

Different approaches have been proposed for solving this problem; we mention among them those using interpolation techniques, γ-iteration approaches, all-pass embedding methods, and, more recently, approaches based on the generalized Popov–Yakubovich theory; corresponding bibliographical details are given in Notes and References.

This chapter is concerned with one and two-block Nehari problems, whose suboptimal solutions are obtained using the generalized Popov–Yakubovich theory; in addition, optimal solutions are also determined via the singular perturbations approach. Illustrative examples including a model matching problem and an optimal robust design with respect to the left coprime factorization are also presented.

Although the two-block Nehari problem is more complex than the one-block case, we shall treat first this problem which will allow us to obtain directly the solution corresponding to the one-block Nehari problem.

4.1. THE TWO-BLOCK CASE

4.1.1. SUBOPTIMAL SOLUTION

Consider the system having the following structure

$$T := \begin{bmatrix} T_1 \\ T_2 \end{bmatrix} = \left[\begin{array}{c|c} A & B \\ \hline C_1 & D_1 \\ C_2 & D_2 \end{array}\right], \tag{4.1}$$

where $A \in \mathbf{R}^{n\times n}$ is *antistable*, $B \in \mathbf{R}^{n\times m}$, $C_i \in \mathbf{R}^{p_i\times n}$, $D_i \in \mathbf{R}^{p_i\times m}$, $i = 1,2$. Then, given $\gamma > 0$, the suboptimal two-block Nehari problem consists in finding a *stable* system S such that

$$\left\| \begin{bmatrix} T_1 + S \\ T_2 \end{bmatrix} \right\|_\infty < \gamma. \tag{4.2}$$

In the following we shall assume that $D_1 = 0$; there is no loss of generality in this assumption since if S is a solution to the Nehari problem with $D_1 = 0$, then $S - D_1$ is a solution to the Nehari problem with arbitrary D_1.

Necessary and sufficient conditions for solvability of the suboptimal two-block Nehari problem, as well as a solution of this problem, are given by the following result:

Theorem 4.1 *The following assertions are true:*

(1) The suboptimal two-block Nehari problem (4.2) has a solution if and only if

$$\gamma^2 I - D_2^T D_2 > 0,$$

and the ARE

$$-AZ - ZA^T + \left[Z(C_2^T D_2 - XB) - B\right]\left(\gamma^2 I - D_2^T D_2\right)^{-1} \times \left[(D_2^T C_2 - B^T X)Z - B^T\right] - ZC_1^T C_1 Z = 0 \quad (4.3)$$

has a stabilizing positive semi-definite solution Z, where $X \geq 0$ is the solution to the Lyapunov equation

$$A^T X + XA - C_1^T C_1 - C_2^T C_2 = 0. \quad (4.4)$$

(2) If the conditions from point (1) hold, then a solution to the suboptimal two-block Nehari problem is given by

$$S = \left[\begin{array}{c|c} -\left(A + ZC_1^T C_1\right)^T & C_2^T D_2 - XB \\ \hline -C_1 Z & 0 \end{array}\right]. \quad (4.5)$$

Proof. (1) *Necessity* Let S be a stable system satisfying (4.2); then the condition $\gamma^2 I - D_2^T D_2 > 0$ follows directly from

$$\|T_2\|_\infty \leq \left\| \begin{bmatrix} T_1 + S \\ T_2 \end{bmatrix} \right\|_\infty < \gamma.$$

Now we prove that (4.3) has a stabilizing positive semi-definite solution. From (4.2) it follows that

$$(T_1 + S)^*(T_1 + S) + T_2^* T_2 < \gamma^2 I,$$

which may be rewritten in the equivalent form

$$\begin{bmatrix} I & S^* \end{bmatrix} \Pi \begin{bmatrix} I \\ S \end{bmatrix} < 0 \quad \text{on} \quad j\overline{\mathbf{R}}, \quad (4.6)$$

where

$$\Pi := \begin{bmatrix} T_1^* T_1 + T_2^* T_2 - \gamma^2 I_m & T_1^* \\ T_1 & I_{p_1} \end{bmatrix}. \tag{4.7}$$

Direct computations show that Π is just the Popov function associated with the Popov triplet

$$\begin{aligned} \overline{\Sigma} &= \left(\overline{A}, \overline{B}; \overline{Q}, \overline{L}, \overline{R}\right) \\ &= \left(A, [B \;\; 0_{n\times p_1}]; C_1^T C_1 + C_2^T C_2, \left[C_2^T D_2 \;\; C_1^T\right],\right. \\ &\quad \left.\begin{bmatrix} -\gamma^2 I_m + D_2^T D_2 & 0 \\ 0 & I_{p_1} \end{bmatrix}\right). \end{aligned}$$

Denote by $\widehat{\Sigma}$ the $(-X, 0_{m\times n})$-equivalent Popov triplet to $\overline{\Sigma}$; then from the definition of the Popov triplets equivalence, we obtain

$$\begin{aligned} \widehat{\Sigma} &= \left(\widehat{A}, \widehat{B}; \widehat{Q}, \widehat{L}, \widehat{R}\right) \\ &= \left(A, [B \;\; 0_{n\times p_1}]; 0_{n\times n}, \left[C_2^T D_2 - XB \;\; C_1^T\right],\right. \\ &\quad \left.\begin{bmatrix} -\gamma^2 I_m + D_2^T D_2 & 0 \\ 0 & I_{p_1} \end{bmatrix}\right). \end{aligned}$$

Consider also the dual $\widetilde{\Sigma}$ of $\widehat{\Sigma}$

$$\begin{aligned} \widetilde{\Sigma} &= \left(\widetilde{A}, \widetilde{B}; \widetilde{Q}, \widetilde{L}, \widetilde{R}\right) \\ &= \left(-A^T, -\left[C_2^T D_2 - XB \;\; C_1^T\right]; 0_{n\times n}, [B \;\; 0_{n\times p_1}],\right. \\ &\quad \left.\begin{bmatrix} -\gamma^2 I_m + D_2^T D_2 & 0 \\ 0 & I_{p_1} \end{bmatrix}\right). \end{aligned}$$

Then from Proposition 2.5 it results that $\Pi_{\widetilde{\Sigma}} = \Pi_{\widehat{\Sigma}}$. Further, since $\widehat{\Sigma}$ is $(-X, 0)$-equivalent to $\overline{\Sigma}$ it follows from (2) of Proposition 2.4 that

$$\Pi = \Pi_{\overline{\Sigma}} = S_{\widehat{F}}^* \Pi_{\widehat{\Sigma}} S_{\widehat{F}} = S_{\widehat{F}}^* \Pi_{\widetilde{\Sigma}} S_{\widehat{F}},$$

where

$$S_{\widehat{F}} = \left[\begin{array}{c|c} A & B \\ \hline 0 & I \end{array}\right] = I.$$

Thus we deduce that

$$\Pi = \Pi_{\overline{\Sigma}} = \Pi_{\widetilde{\Sigma}}. \tag{4.8}$$

Then (4.6) and (4.8) give

$$\begin{bmatrix} I & S^* \end{bmatrix} \Pi_{\widetilde{\Sigma}} \begin{bmatrix} I \\ S \end{bmatrix} < 0 \quad \text{on} \quad j\overline{\mathbf{R}}. \tag{4.9}$$

Writing $\Pi_{\widetilde{\Sigma}}$ in the partitioned form according to (4.7), namely,

$$\Pi_{\widetilde{\Sigma}} = \begin{bmatrix} \Pi_{\widetilde{\Sigma},11} & \Pi_{\widetilde{\Sigma},12} \\ \Pi^*_{\widetilde{\Sigma},12} & \Pi_{\widetilde{\Sigma},22} \end{bmatrix},$$

and taking into account that $\Pi_{\widetilde{\Sigma}} = \Pi$, we obtain using (4.7)

$$\Pi_{\widetilde{\Sigma},22} = I_{p_1} > 0 \quad \text{on} \quad j\overline{\mathbf{R}}.$$

Combining this conclusion with (4.9) and then invoking Theorem 2.8 one concludes that the KPYS$(\widetilde{\Sigma}, J)$ with

$$J = \operatorname{sgn} \widetilde{R} = \begin{bmatrix} -I_m & \\ & I_{p_1} \end{bmatrix}$$

has a stabilizing solution and hence the ARE$(\widetilde{\Sigma})$ (4.3) has a stabilizing solution Z. Furthermore, since

$$\begin{bmatrix} \widetilde{Q} & \widetilde{L}_2 \\ \widetilde{L}_2^T & \widetilde{R}_{22} \end{bmatrix} = \begin{bmatrix} 0 & 0 \\ 0 & I \end{bmatrix} \geq 0,$$

it follows that $Z \geq 0$.

Sufficiency Suppose that $\gamma^2 I - D_2^T D_2 > 0$ and that the ARE$(\widetilde{\Sigma})$ (4.3) has a stabilizing solution $Z \geq 0$. Then, clearly sgn $\widetilde{R} = J$ and the KPYS$(\widetilde{\Sigma},J)$ has a stabilizing solution. Since the KPYS$(\widetilde{\Sigma},J)$ has a solution, according to (2) of Proposition 2.4 it follows that the spectral factorization identity

$$\Pi_{\widetilde{\Sigma}} = S^*_{\widetilde{F}} \widetilde{R} S_{\widetilde{F}} \tag{4.10}$$

hold, where

$$S_{\widetilde{F}} = \left[\begin{array}{c|c} \widetilde{A} & \widetilde{B} \\ \hline -\widetilde{F} & I \end{array}\right]$$

is the spectral factor with both $S_{\widetilde{F}}$ and $S_{\widetilde{F}}^{-1}$ in $\mathrm{RH}^\infty_{+,(m+p_1)\times(m+p_1)}$. According with (2.15)

$$\begin{aligned} \widetilde{F} &:= -\widetilde{R}^{-1}\left(\widetilde{B}^T Z + \widetilde{L}^T\right) = \begin{bmatrix} \widetilde{F}_1 \\ \widetilde{F}_2 \end{bmatrix} \\ &= \begin{bmatrix} -\left(\gamma^2 I - D_2^T D_2\right)^{-1}\left(\left(C_2^T D_2 - XB\right)^T Z - B^T\right) \\ C_1 Z \end{bmatrix}. \end{aligned} \tag{4.11}$$

Further, taking into account that

$$\tilde{R} = V^T J V$$

with

$$V := \begin{bmatrix} \left(\gamma^2 I_m - D_2^T D_2\right)^{\frac{1}{2}} & 0 \\ 0 & I_{p_1} \end{bmatrix} \quad \text{and} \quad J := \begin{bmatrix} -I_m & 0 \\ 0 & I_{p_1} \end{bmatrix},$$

(4.8) and (4.10) give

$$\Pi = G^* J G, \tag{4.12}$$

where

$$G := V S_{\widetilde{F}}.$$

Now define

$$\begin{bmatrix} S_1 \\ S_2 \end{bmatrix} := G^{-1} \begin{bmatrix} I_m \\ 0_{p_1 \times m} \end{bmatrix} \tag{4.13}$$

and note that $G^{-1} \in \mathrm{RH}^{\infty}_{+,(m+p_1)\times(m+p_1)}$ with

$$G^{-1} = \left[\begin{array}{c|c} \tilde{A} + \tilde{B}\widetilde{F} & \tilde{B}V^{-1} \\ \hline \widetilde{F} & V^{-1} \end{array}\right].$$

Then from (4.13) we obtain that

$$\begin{bmatrix} S_1 \\ S_2 \end{bmatrix} = \left[\begin{array}{c|c} \tilde{A} + \tilde{B}\widetilde{F} & -\left(C_2^T D_2 - XB\right)\left(\gamma^2 I - D_2^T D_2\right)^{-\frac{1}{2}} \\ \hline \widetilde{F}_1 & \left(\gamma^2 I - D_2^T D_2\right)^{-\frac{1}{2}} \\ \widetilde{F}_2 & 0 \end{array}\right], \tag{4.14}$$

from which we obtain with (4.11)

$$S_1^{-1} = \left[\begin{array}{c|c} -\left(A + Z C_1^T C_1\right)^T & C_2^T D_2 - XB \\ \hline \left(\gamma^2 I - D_2^T D_2\right)^{\frac{1}{2}} \widetilde{F}_1 & \left(\gamma^2 I - D_2^T D_2\right)^{\frac{1}{2}} \end{array}\right]. \tag{4.15}$$

On the other hand, the ARE($\widetilde{\Sigma}$) (4.3) can be written in the Lyapunov equivalent form

$$\begin{aligned} &- \left(A + Z C_1^T C_1\right) Z - Z \left(A + Z C_1^T C_1\right)^T \\ &+ \left[Z\left(C_2^T D_2 - XB\right) - B\right]\left(\gamma^2 I - D_2^T D_2\right)^{-1} \\ &\quad \times \left[Z\left(C_2^T D_2 - XB\right) - B\right]^T + Z C_1^T C_1 Z = 0. \end{aligned}$$

Since $-A^T$ is stable it follows that the pair $\left(-\left(A^T + C_1^T C_1 Z\right), C_1 Z\right)$ is detectable; then as $Z \geq 0$, the equation above shows that $-\left(A + Z C_1^T C_1\right)$ is stable and therefore S_1^{-1} is stable.

We prove now that $S := S_2 S_1^{-1}$ is a solution to the two-block Nehari problem. First, notice that S is stable because both S_2 and S_1^{-1} are stable; then, using (4.8), (4.12) and (4.13) it results

$$\begin{aligned} 0 > [\, I \;\; 0 \,] J \begin{bmatrix} I \\ 0 \end{bmatrix} &= [\, S_1^* \;\; S_2^* \,] G^* J G \begin{bmatrix} S_1 \\ S_2 \end{bmatrix} \\ &= S_1^* [\, I \;\; S^* \,] \Pi \begin{bmatrix} I \\ S \end{bmatrix} S_1 \quad \text{on } j\mathbf{R} \end{aligned}$$

from which we deduce that (4.6) holds, and hence S is a solution of the two-block Nehari problem.

Direct computations based on (4.14) and (4.15) show after removal of the uncontrollable stable part, that $S = S_2 S_1^{-1}$ coincides with (4.5) and hence, (2) in the statement is also proved. □

Using Theorem 4.1, we may determine alternative necessary and sufficient conditions for solving the two-block Nehari problem, recovering thus the known results proved in (Jonckheere *et al.*, 1989)

Corollary 4.1 *The suboptimal two-block Nehari problem has a solution if and only if the following conditions hold:*

(a)

$$D_2^T D_2 < \gamma^2 I;$$

(b) The ARE

$$\begin{aligned} -AY - YA^T + \left(YC_2^T - BD_2^T\right)\left(\gamma^2 I - D_2 D_2^T\right)^{-1} & \\ \times \left(C_2 Y - D_2 B^T\right) + BB^T = 0 & \end{aligned} \tag{4.16}$$

has a stabilizing solution $Y \geq 0$;

(c)

$$\rho(XY) < \gamma^2, \tag{4.17}$$

where X is the positive semi-definite solution to the Lyapunov equation (4.4).

If the conditions (b) and (c) hold, then (4.3) has a stabilizing solution $Z \geq 0$ and

$$Z = Y\left(\gamma^2 I - XY\right)^{-1}. \tag{4.18}$$

Proof. For the first part of the corollary we shall prove that the conditions (a), (b), (c) in the statement are equivalent to those from point (1) in Theorem 4.1; since (a) is a common condition for both statements it remains to show that (b) and (c) are equivalent to the property of the ARE (4.3) to have a stabilizing positive semi-definite solution.

'ARE (4.3)⇒(b) and (c)'. Since the ARE (4.3) has a stabilizing solution $Z \geq 0$, according to Remark 2.1 and Theorem 2.7 the EHP($\widetilde{\Sigma}$) is regular and stable disconjugate and therefore

$$N_{\widetilde{\Sigma}} V_{\widetilde{\Sigma}} = M_{\widetilde{\Sigma}} V_{\widetilde{\Sigma}} S_{\widetilde{\Sigma}}, \tag{4.19}$$

where

$$\begin{aligned} M_{\widetilde{\Sigma}} &:= \begin{bmatrix} I & 0 & 0 \\ 0 & I & 0 \\ 0 & 0 & 0 \end{bmatrix}, \\ N_{\widetilde{\Sigma}} &:= \begin{bmatrix} \tilde{A} & 0 & \tilde{B} \\ 0 & -\tilde{A}^T & -\tilde{L} \\ \tilde{L}^T & \tilde{B}^T & \tilde{R} \end{bmatrix}, \\ V_{\widetilde{\Sigma}} &:= \begin{bmatrix} I \\ Z \\ \widetilde{F} \end{bmatrix}, \end{aligned} \tag{4.20}$$

$\widetilde{F}$ is defined by (4.11), and $S_{\widetilde{\Sigma}}$ is stable. Then (4.19) may be written in the following equivalent form:

$$N_{\widetilde{\Sigma},e} V_{\widetilde{\Sigma},e} = M_{\widetilde{\Sigma},e} V_{\widetilde{\Sigma},e} S_{\widetilde{\Sigma},e} \tag{4.21}$$

obtained by extending the matrices $M_{\widetilde{\Sigma}}$, $N_{\widetilde{\Sigma}}$ and $V_{\widetilde{\Sigma}}$ as follows:

$$\begin{aligned} M_{\widetilde{\Sigma},e} &:= \begin{bmatrix} I & 0 & 0 & 0 & 0 \\ 0 & I & 0 & 0 & 0 \\ 0 & 0 & 0 & 0 & 0 \\ 0 & 0 & 0 & 0 & 0 \\ 0 & 0 & 0 & 0 & 0 \end{bmatrix}, \\ N_{\widetilde{\Sigma},e} &:= \begin{bmatrix} -A^T & 0 & -(C_2^T D_2 - XB) & -C_1^T & 0 \\ 0 & A & -B & 0 & 0 \\ B^T & -(D_2^T C_2 - B^T X) & -\gamma^2 I + D_2^T D_2 & 0 & 0 \\ 0 & -C_1 & 0 & I_{p_1} & 0 \\ 0 & 0 & 0 & 0 & I_{p_2} \end{bmatrix}, \end{aligned}$$

$$V_{\widetilde{\Sigma},e} := \begin{bmatrix} I \\ Z \\ \widetilde{F}_1 \\ \widetilde{F}_2 \\ 0 \end{bmatrix}. \tag{4.22}$$

Further, let us define

$$U := \begin{bmatrix} I & X & C_2^T D_2(D_2^T D_2 - \gamma^2 I)^{-1} & C_1^T & C_2^T(D_2 D_2^T - \gamma^2 I)^{-1} \\ 0 & \gamma^2 I & \gamma^2 B(D_2^T D_2 - \gamma^2 I)^{-1} & 0 & BD_2^T(D_2 D_2^T - \gamma^2 I)^{-1} \\ 0 & 0 & 0 & 0 & I \\ 0 & 0 & I & 0 & 0 \\ 0 & 0 & 0 & I & 0 \end{bmatrix}, \tag{4.23}$$

and

$$\widetilde{Z} := \begin{bmatrix} I & -\gamma^{-2}X & 0 & 0 & 0 \\ 0 & \gamma^{-2}I & 0 & 0 & 0 \\ -(D_2^T D_2 - \gamma^2 I)^{-1} \times B^T & (D_2^T D_2 - \gamma^2 I)^{-1} \times D_2^T C_2 & 0 & I & 0 \\ 0 & \gamma^{-2} C_1 & 0 & 0 & I \\ -D_2 B^T & C_2 & D_2 D_2^T - \gamma^2 I & 0 & 0 \end{bmatrix} \tag{4.24}$$

which are both nonsingular. Then, pre-multiplying (4.21) by U, one obtains

$$N_{\Sigma,e} V_{\Sigma,e} = M_{\Sigma,e} V_{\Sigma,e} S_{\widetilde{\Sigma}}, \tag{4.25}$$

where

$$\begin{aligned} M_{\Sigma,e} &:= U M_{\widetilde{\Sigma},e} \widetilde{Z} = \begin{bmatrix} I & 0 & 0 & 0 & 0 \\ 0 & I & 0 & 0 & 0 \\ 0 & 0 & 0 & 0 & 0 \\ 0 & 0 & 0 & 0 & 0 \\ 0 & 0 & 0 & 0 & 0 \end{bmatrix}, \\ N_{\Sigma,e} &:= U N_{\widetilde{\Sigma},e} \widetilde{Z} \\ &= \begin{bmatrix} -A^T & 0 & C_2^T & 0 & 0 \\ -BB^T & A & BD_2^T & 0 & 0 \\ -D_2 B^T & C_2 & D_2 D_2^T - \gamma^2 I & 0 & 0 \\ 0 & 0 & 0 & D_2 D_2^T - \gamma^2 I & 0 \\ 0 & 0 & 0 & 0 & I \end{bmatrix}, \quad (4.26) \\ V_{\Sigma,e} &:= \widetilde{Z}^{-1} V_{\widetilde{\Sigma},e} = \begin{bmatrix} I + XZ \\ \gamma^2 Z \\ * \\ * \\ * \end{bmatrix}, \end{aligned}$$

the asterisk denoting irrelevant entries.

Equating the first 3×3 block entries in (4.25) one obtains

$$N_\Sigma V_\Sigma = M_\Sigma V_\Sigma S_{\tilde{\Sigma}}, \tag{4.27}$$

with

$$\begin{aligned} M_\Sigma &:= \begin{bmatrix} I & 0 & 0 \\ 0 & I & 0 \\ 0 & 0 & 0 \end{bmatrix}, \\ N_\Sigma &:= \begin{bmatrix} -A^T & 0 & C_2^T \\ -BB^T & A & BD_2^T \\ -D_2B^T & C_2 & D_2D_2^T - \gamma^2 I \end{bmatrix}, \\ V_\Sigma &:= \begin{bmatrix} I + XZ \\ \gamma^2 Z \\ * \end{bmatrix}. \end{aligned} \tag{4.28}$$

Since $X \geq 0$ and $Z \geq 0$ it follows that $I + XZ > 0$, and then, when post-multiplying (4.27) by $(I + XZ)^{-1}$, it results that

$$N_\Sigma \begin{bmatrix} I \\ \gamma^2 Z(I + XZ)^{-1} \\ * \end{bmatrix} = M_\Sigma \begin{bmatrix} I \\ \gamma^2 Z(I + XZ)^{-1} \\ * \end{bmatrix} S_\Sigma, \tag{4.29}$$

where

$$S_\Sigma := (I + XZ) S_{\tilde{\Sigma}} (I + XZ)^{-1},$$

which is stable since $S_{\tilde{\Sigma}}$ is stable. Then, using Theorem 2.7, we deduce from (4.29) that $\gamma^2 Z(I + XZ)^{-1}$ is the stabilizing solution of the ARE(Σ) for

$$\Sigma := \left(-A^T, C_2^T; BB^T, -BD_2^T, D_2D_2^T - \gamma^2 I\right),$$

which is just (4.16). Since the stabilizing solution of (4.16) is Y and it is unique, one obtains

$$Y = \gamma^2 Z(I + XZ)^{-1}.$$

Using the expression above it follows that

$$\gamma^2 I - XY = \gamma^2 I - \gamma^2 XZ(I + XZ)^{-1} = \gamma^2 (I + XZ)^{-1} > 0,$$

which shows that $\gamma^2 I - XY > 0$ and therefore condition c) in the statement holds.

'(b) and (c)⇒ARE(4.3)'. Now we prove that if conditions (b) and (c) hold then (4.3) has a stabilizing solution $Z \geq 0$. Indeed, owing to condition (b), based on Remark 2.1 and Theorem 2.7, there exists S_Σ stable such that

$$N_\Sigma \widetilde{V}_\Sigma = M_\Sigma \widetilde{V}_\Sigma S_\Sigma, \tag{4.30}$$

where

$$\widetilde{V}_\Sigma := \begin{bmatrix} I \\ Y \\ F \end{bmatrix},$$

M_Σ and N_Σ are defined by (4.28), and $\lambda M_\Sigma - N_\Sigma$ is the EHP(Σ). When extending the EHP(Σ), (4.30) gives

$$N_{\Sigma,e} \widetilde{V}_{\Sigma,e} = M_{\Sigma,e} \widetilde{V}_{\Sigma,e} S_\Sigma \tag{4.31}$$

where $M_{\Sigma,e}$ and $N_{\Sigma,e}$ are defined by (4.26) and

$$\widetilde{V}_{\Sigma,e} := \begin{bmatrix} I \\ Y \\ F \\ 0 \\ 0 \end{bmatrix}.$$

From (4.31) it follows that

$$U^{-1} N_{\Sigma,e} \widetilde{Z}^{-1} \widetilde{Z} \widetilde{V}_{\Sigma,e} = U^{-1} M_{\Sigma,e} \widetilde{Z}^{-1} \widetilde{Z} \widetilde{V}_{\Sigma,e} S_\Sigma, \tag{4.32}$$

with U and $\widetilde{Z}$ given by (4.23) and (4.24) respectively. Then, (4.32) may be written in the equivalent form

$$N_{\widetilde{\Sigma},e} \widetilde{V}_{\widetilde{\Sigma},e} = M_{\widetilde{\Sigma},e} \widetilde{V}_{\widetilde{\Sigma},e} S_\Sigma, \tag{4.33}$$

where $M_{\widetilde{\Sigma},e} = U^{-1} M_{\Sigma,e} \widetilde{Z}^{-1}$, $N_{\widetilde{\Sigma},e} = U^{-1} N_{\Sigma,e} \widetilde{Z}^{-1}$ are just the ones defined by (4.22) (due to (4.26)), and

$$\widetilde{V}_{\widetilde{\Sigma},e} := \widetilde{Z} \widetilde{V}_{\Sigma,e} = \begin{bmatrix} I - \gamma^{-2} XY \\ \gamma^{-2} Y \\ * \\ * \\ * \end{bmatrix}. \tag{4.34}$$

When extracting the first 4×4 block from (4.33), we obtain by using the fact that $I - \gamma^{-2} XY$ is invertible,

$$N_{\widetilde{\Sigma}} \widetilde{V}_{\widetilde{\Sigma}} = M_{\widetilde{\Sigma}} \widetilde{V}_{\widetilde{\Sigma}} \widetilde{S}_\Sigma,$$

with $M_{\widetilde{\Sigma}}$, $N_{\widetilde{\Sigma}}$ defined by (4.20),

$$\widetilde{V}_{\widetilde{\Sigma}} := \begin{bmatrix} I \\ Y(\gamma^2 I - XY)^{-1} \\ * \\ * \end{bmatrix},$$

and

$$\widetilde{S}_{\Sigma} := \left(I - \gamma^{-2}XY\right) S_{\Sigma} \left(I - \gamma^{-2}XY\right)^{-1}$$

which is stable since S_{Σ} is stable. Then, from (4.33) and (4.34) we deduce, using again Theorem 2.7, that the ARE $(\widetilde{\Sigma})$ (4.3) has the stabilizing solution $Y(\gamma^2 I - XY)^{-1}$ which is positive semi-definite since $Y \geq 0$ and $\rho(XY) < \gamma^2$. Moreover, based on uniqueness arguments it follows that Z equals $Y(\gamma^2 I - XY)^{-1}$ and hence the proof ends. □

In the next section we shall give an optimal solution to the two-block Nehari problem, obtained via the singular perturbation method.

4.1.2. AN OPTIMAL SOLUTION OF THE TWO-BLOCK NEHARI PROBLEM

Consider the optimal two-block Nehari problem which consists in determining

$$\inf_{S \in \mathrm{RH}_{+}^{\infty}} \left\| \begin{bmatrix} T_1 + S \\ T_2 \end{bmatrix} \right\|_{\infty} := \gamma_0, \tag{4.35}$$

where T_1, $T_2 \in \mathrm{RH}_{-}^{\infty}$ are given by (4.1). Such an optimization problem is important for applications in which the minimum level of attenuation is required.

The following result will be useful for the further developments.

Lemma 4.1 *The function $\gamma \mapsto \rho(XY(\gamma))$ is non-increasing.*

Proof. Let $\gamma_1 > \gamma_2 > 0$ for which the conditions (a), (b) and (c) in Corollary 4.1 hold and denote by $Y(\gamma_1)$ and $Y(\gamma_2)$ the stabilizing conditions corresponding to the Riccati equation (4.16), namely

$$\begin{aligned} -AY(\gamma_1) - Y(\gamma_1)A^T + \left[Y(\gamma_1)C_2^T - BD_2^T\right]\left(\gamma_1^2 I - D_2 D_2^T\right)^{-1} \\ \times \left[C_2 Y(\gamma_1) - D_2 B^T\right] + BB^T = 0 \\ -AY(\gamma_2) - Y(\gamma_2)A^T + \left[Y(\gamma_2)C_2^T - BD_2^T\right]\left(\gamma_2^2 I - D_2 D_2^T\right)^{-1} \\ \times \left[C_2 Y(\gamma_2) - D_2 B^T\right] + BB^T = 0. \end{aligned}$$

Subtracting the two equations above, it results that

$$\begin{aligned}
&- \left\{A - \left[Y(\gamma_1)C_2^T - BD_2^T\right]\left(\gamma_1^2 I - D_2D_2^T\right)^{-1} C_2\right\}[Y(\gamma_1) - Y(\gamma_2)] \\
&- [Y(\gamma_1) - Y(\gamma_2)]\left\{A - \left[Y(\gamma_1)C_2^T - BD_2^T\right]\left(\gamma_1^2 I - D_2D_2^T\right)^{-1} C_2\right\}^T \\
&- [Y(\gamma_1) - Y(\gamma_2)]\, C_2^T \left(\gamma_1^2 I - D_2D_2^T\right)^{-1} C_2\, [Y(\gamma_1) - Y(\gamma_2)] \\
&- \left[Y(\gamma_2)C_2^T - BD_2^T\right]\left[\left(\gamma_2^2 I - D_2D_2^T\right)^{-1} - \left(\gamma_1^2 I - D_2D_2^T\right)^{-1}\right] \\
&\qquad\qquad \times \left[C_2Y(\gamma_2) - D_2B^T\right] = 0. \qquad (4.36)
\end{aligned}$$

Since $\gamma_1 > \gamma_2$, and $-\left\{A - \left[Y(\gamma_1)C_2^T - BD_2^T\right]\left(\gamma_1^2 I - D_2D_2^T\right)^{-1} C_2\right\}$ is stable $Y(\gamma_1)$, being the stabilizing solution of (4.16) for $\gamma = \gamma_1$, we deduce from (4.36) that $Y(\gamma_1) - Y(\gamma_2) \leq 0$ and hence, $\rho(XY(\gamma_1)) \leq \rho(XY(\gamma_2))$.□

Based on Corollary 4.1 and on the lemma above, we obtain the following result:

Theorem 4.2 *The optimum γ_0 is equal to:*

(a) The unique solution to the transcendental equation

$$\gamma^2 = \rho\,(XY(\gamma)) \qquad (4.37)$$

if such a solution exists; or

(b) $\| T_2 \|_\infty$ if (4.37) has no solution.

Proof. First, notice that (4.16) is just the norm type of Riccati equation associated to the system $\left(-A^T, C_2^T, -B^T, -D_2^T\right)$ whose H^∞-norm is equal to the L^∞-norm of T_2. Therefore the Bounded Real Lemma (Corollary 2.7) shows that $\gamma > \|T_2\|_\infty$.

Let us assume that (4.37) has a solution; this solution is unique since the function $\gamma^2 - \rho(XY(\gamma))$ is continuous and increasing. Denote by $\widehat{\gamma}$ the solution of this equation. If $\gamma_0 < \widehat{\gamma}$, then for a $\tilde{\gamma}$ such that $\gamma_0 < \tilde{\gamma} < \widehat{\gamma}$ we have $\tilde{\gamma}^2 < \rho\,(XY(\tilde{\gamma}))$, and hence, according to Corollary 4.1, the suboptimal Nehari problem has no solution for $\tilde{\gamma}$; therefore, γ_0 cannot be less than $\hat{\gamma}$. On the other hand, the suboptimal Nehari problem has a solution for any $\gamma > \hat{\gamma}$, since all conditions in the statement of Corollary 4.1 hold. Then, taking into account that $\gamma_0 \geq \hat{\gamma}$, as we have shown above, we conclude that the minimal Nehari distance γ_0 coincides with the unique solution $\hat{\gamma}$ of (4.37).

Consider now the case when (4.37) has no solution. Since in this case $\gamma^2 - \rho(XY(\gamma)) > 0$ for all $\gamma > \|T_2\|_\infty$ it follows that the suboptimal Nehari problem has a solution for all $\gamma > \|T_2\|_\infty$, from which we conclude that in this case, $\gamma_0 = \|T_2\|_\infty$. □

Based on Corollary 4.1 and Theorem 4.2 we obtain the following γ-procedure to compute the *optimal Nehari distance* γ_0 within an imposed level of tolerance $\epsilon > 0$:

Step 1 Set $\gamma_L := \| T_2 \|_\infty$ and $\gamma = \gamma_L + \epsilon$; if $\gamma^2 - \rho(XY(\gamma)) > 0$, then $\gamma_0 = \gamma$ and STOP; otherwise, go to Step 2;

Step 2 Find $\gamma_U > \| T_2 \|_\infty$ such that $\gamma_U^2 - \rho(XY(\gamma_U)) > 0$ and go to Step 3;

Step 3 If $\gamma_U - \gamma_L < \epsilon$, then $\gamma_0 = \gamma_U$ and STOP; otherwise set $\gamma = (\gamma_L + \gamma_U)/2$ and go to Step 4;

Step 4 If $\gamma^2 - \rho(XY(\gamma)) > 0$, then set $\gamma_L = \gamma$ and return to Step 3.

Consider now the case when the optimum γ_0 verifies (4.37); then the computation of an optimal solution to the Nehari problem, using (4.5) and (4.18) is ill-conditioned since the matrix $\gamma^2 I - XY(\gamma)$ tends to become singular for γ close to γ_0. In the following, we shall give a procedure based on the singular perturbations method, in order to avoid this ill-conditioning near the optimum.

Assume that the realization in (4.1) is minimal and balanced with respect to X and Y for $\gamma > \gamma_0$, that is, X and Y are diagonal and equal. Notice that such a realization may be always obtained from an arbitrary minimal realization (A, B, C, D) of $\begin{bmatrix} T_1 \\ T_2 \end{bmatrix}$ by performing the following procedure, similar to the one described in Section 1.4 for the balancing with respect to the controllability and observability Gramians:

Step 1 Compute the solutions X and $Y(\gamma)$ of (4.4) and (4.16), respectively;

Step 2 Perform a Cholesky factorization $X = \widehat{Z}^T \widehat{Z}$;

Step 3 Determine the singular value descomposition

$$\widehat{Z} Y(\gamma) \widehat{Z}^T = U(\gamma) \Sigma^2(\gamma) U^T(\gamma)$$

with $U(\gamma)$ orthogonal;

Step 4 With the transformation $T_b(\gamma) := \Sigma^{-\frac{1}{2}}(\gamma) U^T(\gamma) \widehat{Z}$, determine the equivalent realization

$$\left(T_b(\gamma) A T_b^{-1}(\gamma),\ T_b(\gamma) B,\ C T_b^{-1}(\gamma),\ D \right)$$

which is balanced with respect to X and Y.

The proof that the equivalent realization obtained by the procedure above is balanced with respect to X and Y is straightforward, and therefore it is omitted.

By an abuse of notation the same letters A, B and C will be used for the balanced realization of the system; the matrices A, B and C depend on γ and the solutions X and Y are given by

$$X(\gamma) = Y(\gamma) = \begin{bmatrix} r_1(\gamma)I_1 & 0 \\ 0 & R_{22}(\gamma) \end{bmatrix}, \tag{4.38}$$

where $R_{22}(\gamma) = \operatorname{diag}(r_2(\gamma)I_2, ..., r_p(\gamma)I_p)$, with $r_1(\gamma) > ... > r_p(\gamma) > 0$ and I_k are $n_k \times n_k$ unit matrices, $k = 1, ..., p$.

Let $\gamma = \sqrt{\gamma_0^2 + \epsilon}$ with $\epsilon > 0$; since γ_0 verifies (4.37), it follows that $r_1(\gamma) \longrightarrow \gamma_0$ when $\epsilon \longrightarrow 0$ and then, from (4.18) and (4.38) we obtain the following expression for $Z(\gamma)$ in the neighborhood of the optimum

$$Z(\gamma) = \begin{bmatrix} \frac{r_1(\gamma)}{\epsilon} I_1 & 0 \\ 0 & Z_{22}(\gamma) \end{bmatrix}, \tag{4.39}$$

where

$$Z_{22} := R_{22}(\gamma)\left[\gamma^2 I - R_{22}^2(\gamma)\right]^{-1}.$$

According to Theorem 4.1, a suboptimal solution to the two-block Nehari problem is given by (4.5) which gives together with (4.39)

$$S = \left[\begin{array}{c|c} A_s^\epsilon & B_s^\epsilon \\ \hline C_s^\epsilon & 0 \end{array}\right], \tag{4.40}$$

where

$$\begin{aligned} A_s^\epsilon &:= -(A + ZC_1^T C_1)^T = \begin{bmatrix} -A_{11}^T - \frac{r_1}{\epsilon} C_{11}^T C_{11} & -A_{21}^T - C_{11}^T C_{12} Z_{22} \\ -A_{12}^T - \frac{r_1}{\epsilon} C_{12}^T C_{11} & -A_{22}^T - C_{12}^T C_{12} Z_{22} \end{bmatrix}, \\ B_s^\epsilon &:= C_2^T D_2 - XB = \begin{bmatrix} -r_1 B_1 + C_{21}^T D_2 \\ -R_{22} B_2 + C_{22}^T D_2 \end{bmatrix}, \\ C_s^\epsilon &:= -C_1 Z = -\begin{bmatrix} \frac{r_1}{\epsilon} C_{11} & C_{12} Z_{22} \end{bmatrix}, \end{aligned} \tag{4.41}$$

with A_{ij}, B_i and C_{ij}, $i, j = 1, 2$ determined by the partitions of A, B and C respectively, conformably to the partition (4.39) of Z, namely

$$A = \begin{bmatrix} A_{11} & A_{12} \\ A_{21} & A_{22} \end{bmatrix}, \quad B = \begin{bmatrix} B_1 \\ B_2 \end{bmatrix}, \quad C = \begin{bmatrix} C_{11} & C_{12} \\ C_{21} & C_{22} \end{bmatrix}.$$

Remark 4.1 All matrices A_{ij}, B_i and C_i, $i,j = 1,2$ defined above depend on γ, as well as r_1 and, since $\gamma = \sqrt{\gamma_0^2 + \epsilon}$, they are in fact functions of ϵ; these functions are continuous and have finite limits for $\epsilon \longrightarrow 0$ since the dependence $Y(\gamma)$ is smooth around γ_0. □

Consider now the equivalent realization of S in (4.40) given by the transformation $T_\epsilon := \begin{bmatrix} \frac{1}{\epsilon}I & 0 \\ 0 & I \end{bmatrix}$, namely $(T_\epsilon A_s^\epsilon T_\epsilon^{-1}, T_e B_s^\epsilon, C_s^\epsilon T_\epsilon^{-1})$, to which correspond the following state space equations:

$$\begin{aligned} \epsilon\dot{x}_1 &= -\left(\epsilon A_{11}^T + r_1 C_{11}^T C_{11}\right) x_1 - \left(A_{21}^T + C_{11}^T C_{12} Z_{22}\right) x_2 \\ &\quad + \left(-r_1 B_1 + C_{21}^T D_2\right) u, \\ \dot{x}_2 &= -\left(\epsilon A_{12}^T + r_1 C_{12}^T C_{11}\right) x_1 - \left(A_{22}^T + C_{12}^T C_{12} Z_{22}\right) x_2 \\ &\quad + \left(-R_{22} B_2 + C_{22}^T D_2\right) u, \\ y &= -r_1 C_{11} x_1 - C_{12} Z_{22} x_2. \end{aligned} \tag{4.42}$$

The equations above show that a suboptimal solution of the Nehari problem near the optimum assuming that it verifies (4.37), is in fact a singularly perturbed system. If $C_{11}^T C_{11} \geq \mu I$, with $\mu > 0$ not depending on ϵ for $\epsilon > 0$ small enough, the system (4.42) may be reduced according to singular perturbations theory (Saksena *et al.*, 1984), obtaining thus the following reduced order system:

$$S_0 = \left[\begin{array}{c|c} A_0 & B_0 \\ \hline C_0 & D_0 \end{array}\right], \tag{4.43}$$

with

$$\begin{aligned} A_0 &:= \overline{C}_{12}^T C_0 - \overline{A}_{22}^T, \\ B_0 &:= \overline{C}_{12}^T D_0 - \overline{R}_{22}\overline{B}_2 + \overline{C}_{22}^T \overline{D}_2, \\ C_0 &:= \overline{C}_{11}\left(\overline{C}_{11}^T \overline{C}_{11}\right)^{-1}\left(\overline{A}_{21}^T + \overline{C}_{11}^T \overline{C}_{12}\overline{Z}_{22}\right) - \overline{C}_{12}\overline{Z}_{22}, \\ D_0 &:= \overline{C}_{11}\left(\overline{C}_{11}^T \overline{C}_{11}\right)^{-1}\left(\gamma_0 \overline{B}_1 - \overline{C}_{21}^T \overline{D}_2\right), \end{aligned} \tag{4.44}$$

where the overlined matrices above denote the limits of their corresponding functions when $\epsilon \longrightarrow 0$ by $\epsilon > 0$.

The main result of this subsection, providing an optimal solution to the two-block Nehari problem, is the following theorem:

Theorem 4.3 *Assume that the optimal Nehari distance γ_0 verifies (4.37) and that the realization in (4.1) is balanced with respect to the solutions*

X and Y of (4.4) and (4.16), respectively, corresponding to $\gamma = \gamma_0$; if $\overline{C}_{11}^T\overline{C}_{11}$ is nonsingular, then the system (4.43) is an optimal solution to the two-block Nehari problem.

Proof. We shall show first that the matrix A_0 defined by (4.44) is stable. When multiplying the Riccati equation (4.3) on the left by T_ϵ^{-1} and on the right by T_ϵ, we obtain from its partition the following equations:

- The block (1,1)

$$\begin{aligned} &-\epsilon r_1 A_{11} - r_1^2 C_{11}^T C_{11} - \epsilon r_1 A_{11}^T \\ &+ \left(-\gamma^2 B_1 + r_1 C_{21}^T D_2\right)\left(\gamma^2 I - D_2^T D_2\right)^{-1} \\ &\quad \times \left(-\gamma^2 B_1^T + r_1 D_2^T C_{21}\right) = 0; \end{aligned} \tag{4.45}$$

- The block (1,2)

$$\begin{aligned} &-r_1\left(A_{21}^T + C_{11}^T C_{12} Z_{22}\right) \\ &+ \left(-\gamma^2 B_1 + r_1 C_{21}^T D_2\right)\left(\gamma^2 I - D_2^T D_2\right)^{-1} \\ &\quad \times \left(-\gamma^2 B_2^T + D_2^T C_{22} R_{22}\right)\left(\gamma^2 I - R_{22}^2\right)^{-1} = 0; \end{aligned} \tag{4.46}$$

- The block (2,2)

$$\begin{aligned} &-\left(A_{22} + Z_{22} C_{12}^T C_{12}\right) Z_{22} - Z_{22}\left(A_{22} + Z_{22} C_{12}^T C_{12}\right)^T + Z_{22} C_{12}^T C_{12} Z_{22} \\ &+ \left(\gamma^2 I - R_{22}^2\right)^{-1}\left(-\gamma^2 B_2 + R_{22} C_{22}^T D_2\right)\left(\gamma^2 I - D_2^T D_2\right)^{-1} \\ &\quad \times \left(-\gamma^2 B_2^T + D_2^T C_{22} R_{22}\right)\left(\gamma^2 I - R_{22}^2\right)^{-1} = 0. \end{aligned} \tag{4.47}$$

By substituting (4.45) and (4.46) in (4.47), after some algebraic manipulations we obtain for $\epsilon \longrightarrow 0$ with $\epsilon > 0$ the following Lyapunov inequality:

$$A_0^T \overline{Z}_{22} + \overline{Z}_{22} A_0 + C_0^T C_0 \leq 0. \tag{4.48}$$

On the other hand, from (4.44) it results that $A_0 - \overline{C}_{12}^T C_0 = -\overline{A}_{22}^T$, where $\overline{A}_{22}$ is antistable since the realization (4.1) has been assumed balanced with respect to the solutions of equations (4.4) and (4.16), respectively. This last affirmation is a consequence of Theorem 1.7; indeed, one can directly check that the Gramians of the system

$$\left(-A, \left[\begin{array}{cc} B & \left(YC_2^T - BD_2^T\right)\left(\gamma^2 I - D_2 D_2^T\right)^{-\frac{1}{2}} \end{array}\right], \left[\begin{array}{c} C_1 \\ C_2 \end{array}\right]\right),$$

coincide with the solutions X and Y of (4.4) and (4.16), respectively and since X and Y are diagonal and equal, it follows that the system above is balanced with respect to the Gramians, and according to the first part of Theorem 1.7, $-\overline{A}_{22}$ is stable, and hence $\overline{A}_{22}$ is antistable.

Since $-\overline{A}_{22}$ is stable it results that the pair (C_0, A_0) is detectable and therefore, based on that $\overline{Z}_{22} > 0$, we conclude from the Lyapunov inequality (4.48), that A_0 is stable.

We prove now that S_0 gives the minimal Nehari distance γ_0. To this end consider the system defined as

$$\overline{T} := \begin{bmatrix} T_1 + S_0 \\ T_2 \end{bmatrix} = \left[\begin{array}{c|c} A_T & B_T \\ \hline C_T & D_T \end{array}\right], \tag{4.49}$$

where

$$A_T := \begin{bmatrix} A & 0 \\ 0 & A_0 \end{bmatrix}, \; B_T := \begin{bmatrix} B \\ B_0 \end{bmatrix}, \; C_T := \begin{bmatrix} C_1 & C_0 \\ C_2 & 0 \end{bmatrix}, \; D_T := \begin{bmatrix} D_0 \\ D_2 \end{bmatrix}.$$

Using (4.4), (4.44), (4.45) and (4.47), direct algebraic computations show that for all $\gamma > \gamma_0$, $\gamma^2 I - D_T^T D_T > 0$ and the Riccati inequality

$$\begin{aligned} & A_T^T\Pi + \Pi A_T + \left(\Pi B_T + C_T^T D_T\right)\left(\gamma^2 I - D_T^T D_T\right)^{-1}\left(B_T^T\Pi + D_T^T C_T\right) \\ & \quad + C_T^T C_T \leq 0 \end{aligned} \tag{4.50}$$

is verified by

$$\Pi := \begin{bmatrix} X & \begin{bmatrix} 0 \\ I \end{bmatrix} \\ \begin{bmatrix} 0 & I \end{bmatrix} & -\overline{Z}_{22} \end{bmatrix}.$$

Further, consider the state space equations of the system $\overline{T}$

$$\begin{aligned} \dot{x} &= A_T x + B_T u, \\ y &= C_T x + D_T u. \end{aligned}$$

Then using (4.50) we have for an arbitrary $u \in \mathrm{L}^{2,m}$

$$\begin{aligned} \int_{-\infty}^{\infty} y^T y \, dt &= \int_{-\infty}^{\infty} \left(x^T C_T^T + u^T D_T^T\right)(C_T x + D_T u) \, dt \\ &\leq \int_{-\infty}^{\infty} \Big\{ -x^T \Big[A_T^T\Pi + \Pi A_T + \left(\Pi B_T + C_T^T D_T\right) \\ &\qquad \times \left(\gamma^2 I - D_T^T D_T\right)^{-1} \left(B_T^T \Pi + D_T^T C_T\right)\Big] x \\ &\qquad + u^T D_T^T C_T x + x^T C_T^T D_T u \Big\} \, dt \end{aligned}$$

$$
\begin{aligned}
&= -\int_{-\infty}^{\infty} \frac{d}{dt}\left(x^T \Pi x\right) dt \\
&\quad - \int_{-\infty}^{\infty} \left[x^T \left(\Pi B_T + C_T^T D_T\right)\left(\gamma^2 I - D_T^T D_T\right)^{-1} - u^T\right] \\
&\qquad \times \left(\gamma^2 I - D_T^T D_T\right)\left[\left(\gamma^2 I - D_T^T D_T\right)^{-1}\right. \\
&\qquad \times \left.\left(B_T^T \Pi + D_T^T C_T\right) x - u\right] dt + \gamma^2 \int_{-\infty}^{\infty} u^T u \, dt.
\end{aligned}
$$

Since A_T is dichotomous, the first term in the right side above vanishes; therefore, we obtain that

$$
\int_{-\infty}^{\infty} y^T y \, dt \le \gamma^2 \int_{-\infty}^{\infty} u^T u \, dt
$$

for all $u \in \mathrm{L}^{2,m}$, which shows that $\| \overline{T} \|_\infty \le \gamma$ for all $\gamma \ge \gamma_0$. On the other hand, $\| \overline{T} \|_\infty \ge \gamma_0$ since γ_0 is the minimal Nehari distance; from these two last inequalities we deduce that $\| \overline{T} \|_\infty = \gamma_0$ and hence, S_0 is an optimal solution of the two-block Nehari problem. □

4.1.3. CASE STUDIES

For the study cases described in this section it is useful to present firstly some considerations concerning the coprime factorizations. Let G be a system with the realization (A, B, C). Then the matrices M, N, $\widetilde{M}$, $\widetilde{N}$, U, V, $\widetilde{U}$, $\widetilde{V} \in \mathrm{RH}_+^\infty$ with M and $\widetilde{M}$ invertible, satisfying the conditions

$$
G = NM^{-1} = \widetilde{M}^{-1}\widetilde{N} \tag{4.51}
$$

and

$$
\begin{bmatrix} \widetilde{V} & \widetilde{U} \\ -\widetilde{N} & \widetilde{M} \end{bmatrix} \begin{bmatrix} M & -U \\ N & V \end{bmatrix} = \begin{bmatrix} I & 0 \\ 0 & I \end{bmatrix} \tag{4.52}
$$

define a *double coprime factorization of* G.

Assume that (A, B) is stabilizable and (C, A) is detectable and let F and H be such that $A+BF$ and $A+HC$ are stable. Then a double coprime factorization of T is given by

$$
\begin{aligned}
M &= \left[\begin{array}{c|c} A+BF & B \\ \hline F & I \end{array}\right], & \widetilde{M} &= \left[\begin{array}{c|c} A+HC & H \\ \hline C & I \end{array}\right], \\
N &= \left[\begin{array}{c|c} A+BF & B \\ \hline C & 0 \end{array}\right], & \widetilde{N} &= \left[\begin{array}{c|c} A+HC & B \\ \hline C & 0 \end{array}\right], \\
U &= \left[\begin{array}{c|c} A+BF & H \\ \hline F & 0 \end{array}\right], & \widetilde{U} &= \left[\begin{array}{c|c} A+HC & H \\ \hline F & 0 \end{array}\right], \\
V &= \left[\begin{array}{c|c} A+BF & H \\ \hline -C & I \end{array}\right], & \widetilde{V} &= \left[\begin{array}{c|c} A+HC & B \\ \hline -F & I \end{array}\right],
\end{aligned} \tag{4.53}
$$

as can be immediately checked.

If M, N, $\widetilde{M}$, and $\widetilde{N}$ satisfy the additional conditions

$$M^*M + N^*N = I \tag{4.54}$$

and

$$\widetilde{M}\widetilde{M}^* + \widetilde{N}\widetilde{N}^* = I \tag{4.55}$$

the double coprime factorization is called *normalized.* In order to determine a normalized coprime factorization consider the AREs

$$A^TX + XA - XBB^TX + C^TC = 0 \tag{4.56}$$

and

$$AY + YA^T - YC^TCY + BB^T = 0 \tag{4.57}$$

which have the positive semi-definite stabilizing solutions X and Y, respectively. Then define the Popov triplet

$$\Sigma_1 := \left(A, B; C^TC, 0, I\right).$$

Direct computations show that its associate Popov function has the expression

$$\Pi_\Sigma = G^*G + I. \tag{4.58}$$

According to the spectral factorization identity (2.22) we obtain with

$$S_F = \left[\begin{array}{c|c} A & B \\ \hline B^TX & I \end{array}\right], \quad S_F^{-1} = \left[\begin{array}{c|c} A - BB^TX & B \\ \hline -B^TX & I \end{array}\right] \tag{4.59}$$

where S_F^{-1} is stable since X is the stabilizing solution of (4.56). By pre-multiplying and post-multiplying (4.58) by $\left(S_F^{-1}\right)^*$ and S_F^{-1} respectively, it results that

$$\left(S_F^{-1}\right)^* G^*GS_F^{-1} + \left(S_F^{-1}\right)^* S_F^{-1} = I,$$

from which we deduce that M and N defined as

$$M := S_F^{-1}, \quad N = GS_F^{-1}$$

satisfy (4.54). From the definition of M and from (4.59) it follows that

$$M = \left[\begin{array}{c|c} A - BB^TX & B \\ \hline -B^TX & I \end{array}\right]. \tag{4.60}$$

Using the realizations of G and S_F^{-1} it is easy to see that the state space realization of N defined above contains an unobservable part which-on removal leads to

$$N = \left[\begin{array}{c|c} A - BB^T X & B \\ \hline C & 0 \end{array}\right]. \tag{4.61}$$

Similarily, by considering the Popov triplet

$$\Sigma_2 = \left(-A^T, C^T; BB^T, 0, I\right),$$

one obtains

$$\widetilde{M} = \left[\begin{array}{c|c} A - YC^T C & -YC^T \\ \hline C & I \end{array}\right], \quad \tilde{N} = \left[\begin{array}{c|c} A - YC^T C & B \\ \hline C & 0 \end{array}\right] \tag{4.62}$$

which satisfy (4.55).

The coprime factors $U, V, \widetilde{U}, \widetilde{V}$ corresponding to the normalized case are determined using the formulae (4.53) in which $F = -B^T X$ and $H = -YC^T$.

Case study 1: An optimal two-block H^∞ problem

Consider the optimal two-block H^∞ problem

$$\inf_{K \in \mathrm{RH}_+^\infty} \left\| \begin{bmatrix} R_1(s) + \phi(s)K(s) \\ R_2(s) \end{bmatrix} \right\|_\infty =: \gamma_0 \tag{4.63}$$

where $R_1(s)$, $R_2(s) \in \mathrm{RH}_+^\infty$ and $\phi(s)$ is an inner square transfer function. The problem originates in a mixed sensitivity problem transformed via Youla–Jabr–Bongiorno controller parameterization and it has been also considered in (Chang *et al.*, 1989) to illustrate an alternative algorithm to compute γ_0. In order to reduce this problem to a two-block Nehari problem, we take into account that

$$\begin{aligned} \left\| \begin{bmatrix} R_1 + \phi K \\ R_2 \end{bmatrix} \right\|_\infty &= \left\| \begin{bmatrix} R_1 \\ R_2 \end{bmatrix} + \phi \begin{bmatrix} K \\ 0 \end{bmatrix} \right\|_\infty \\ &= \left\| \phi^* \begin{bmatrix} R_1 \\ R_2 \end{bmatrix} + \begin{bmatrix} K \\ 0 \end{bmatrix} \right\|_\infty . \end{aligned} \tag{4.64}$$

Further, perform the decomposition

$$\phi^* R_1 = P_{1a} + P_{1s} \tag{4.65}$$

with P_{1s} and P_{1a}, stable and antistable, respectively, and the factorization

$$\phi^* R_2 = M_2^{-1} N_2 \tag{4.66}$$

with M_2, N_2 antistable and $M_2^* M_2 = I$. The above factorization may always be determined when starting from a minimal realization (A_2, B_2, C_2, D_2) of $\phi^* R_2$ with A_2 dichotomous. Indeed, one can directly check that M_2 and N_2 given by

$$M_2 = \left[\begin{array}{c|c} A_2 + B_2 F_2 & B_2 \\ \hline F_2 & I \end{array} \right]$$

and

$$N_2 = \left[\begin{array}{c|c} A_2 + B_2 F_2 & B_2 \\ \hline C_2 + D_2 F_2 & D_2 \end{array} \right],$$

respectively, where $F_2 := -B_2^T X_2$ and X_2 denoting the stabilizing solution of the Bernoulli equation

$$A_2^T X_2 + X_2 A_2 - X_2 B_2 B_2^T X_2 = 0\,,$$

satisfy the conditions imposed for the factorization (4.66).

Since $M_2^* M_2 = I$, by (4.64), (4.65) and (4.66) one obtains

$$\left\| \left[\begin{array}{c} R_1 + \phi K \\ R_2 \end{array} \right] \right\|_\infty = \left\| \left[\begin{array}{c} P_{1a} + (P_{1s} + K) \\ N_2 \end{array} \right] \right\|_\infty ,$$

and therefore we have reduced the problem (4.63) to an optimal two-block Nehari problem with $T_1 = P_{1a}$ and $T_2 = N_2$ which solution $S = P_{1s} + K$ provides the solution $K = S - P_{1s}$ of (4.63).

In (Chang *et al.*, 1989) the following numerical example has been considered

$$\begin{aligned} R_1(s) &= \frac{-2(s+10)(s+0.125)(s-0.12)}{(s+0.1)(s+1)(10s+\sqrt{2})} \\ R_2(s) &= \frac{0.1s+1}{10s+\sqrt{2}} \\ \phi(s) &= \frac{(10s-\sqrt{2})(s-1)}{(10s+\sqrt{2})(s+1)}. \end{aligned}$$

Based on the procedure described above, we have transformed the optimization problem in a two-block Nehari problem and using the γ-procedure given in Section 4.1.2 we computed, with a tolerance level $\epsilon = 10^{-12}$, the minimal Nehari distance $\gamma_0 = 1.100437963947$, which coincides with the one obtained in (Chang *et al.*, 1989) by a different algorithm. Then, using formulae (4.44) we determined the optimal solution to the Nehari problem which provided the optimal reduced order controller $K = \left[\begin{array}{c|c} A_K & B_K \\ \hline C_K & D_K \end{array} \right]$,

where

$$\begin{aligned} A_K &= \begin{bmatrix} -0.2039 & -0.0104 \\ 1 & 0 \end{bmatrix}, \quad B_K = \begin{bmatrix} 1 \\ 0 \end{bmatrix} \\ C_K &= \begin{bmatrix} -0.0469 & -0.0049 \end{bmatrix}, \quad D_K = -0.9003. \end{aligned}$$

Study case 2: A model-matching problem

The design problem considered in the following consists in determining a stability augmentation system (SAS) for the short period dynamics of a fighter in order to improve its maneuverability performance. Different design methods are used for solving this problem (*e.g.* (Franklin and Ackermann, 1981), (McLean, 1991), (Sparks and Banda, 1993); in the present case study, a model matching based technique is adopted.

The short period motion of the aircraft including the actuator dynamics is approximated by the third-order linear system

$$\begin{bmatrix} \dot{\alpha} \\ \dot{q} \\ \dot{\delta}_e \end{bmatrix} = \begin{bmatrix} Z_\alpha & 1 & Z_\delta \\ M_\alpha & M_q & M_\delta \\ 0 & 0 & -1/\tau \end{bmatrix} \begin{bmatrix} \alpha \\ q \\ \delta_e \end{bmatrix} + \begin{bmatrix} 0 \\ 0 \\ 1/\tau \end{bmatrix} u, \tag{4.67}$$

with the states α the angle of attack, q the pitch rate, δ_e the elevator deflection and with the control u denoting the coupled commanded deflections of elevator and canard. The stability and the control derivatives Z_α, M_α, M_q, Z_δ, M_δ depend on the flight conditions which are assumed fixed in this case study. The output vector is $[\ a_z \quad q\ ]^T$ where a_z denotes the normal acceleration

$$a_z = \frac{V(\dot{\alpha} - q)}{g} = n_\alpha \alpha + n_\delta \delta_e\,,$$

V denoting the airspeed, and n_α, n_δ are constant coefficients for the nominal flight conditions considered.

In order to accomplish the maneuverability requirements of the aircraft, an 'ideal model' of the short period is first determined, according to the specifications given in (MIL-STD-1797A, 1990); this ideal model has the transfer function

$$H_m(s) = \frac{\omega_m^2}{s^2 + 2\xi_m \omega_m s + \omega_m^2}. \tag{4.68}$$

The design procedure of the SAS is based on the control configuration shown in Figure 4.1, where G denotes the short period dynamics (4.67), α_{ref} is the reference input for the angle of attack, V, W, W_n are weighting functions penalizing the control, the tracking error and the measurement noises n_a and n_q, respectively.

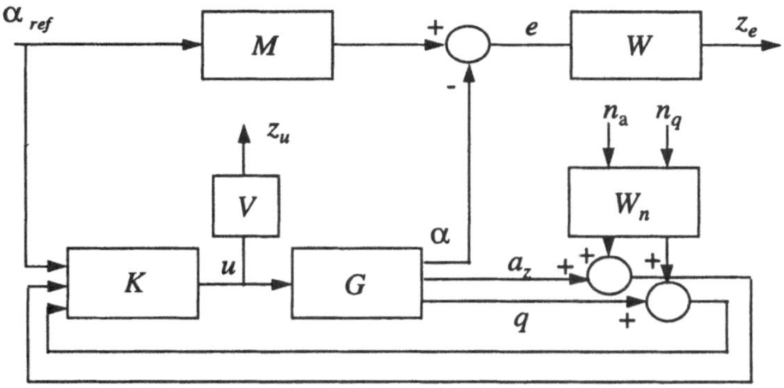

Figure 4.1. **The model-matching control configuration**

The design problem consists in determining K such that the following objectives are achieved:

— The difference between the angle of attack in the short period and the output of the ideal model is minimized.

— The control is limited in order to avoid the actuator saturation.

— The influence of the measurement noises over the tracking error is reduced.

The control configuration shown in Figure 4.1 is equivalent with the H^∞ -control problem illustrated in Figure 3.1, where

$$\begin{aligned} u_1 &= \begin{bmatrix} \alpha_{ref} & n_a & q \end{bmatrix}^T, \quad u_2 = u, \\ y_1 &= \begin{bmatrix} z_e & z_u \end{bmatrix}^T, \quad y_2 = \begin{bmatrix} \alpha_{ref} & a_z + n_a & q + n_q \end{bmatrix}^T \end{aligned}$$

and the realization of the generalized system T will be determined below. Denoting by $T_{y_1u_1}$ the transfer function matrix from u_1 to y_1, it follows that the problem amounts to determine a stabilizing controller K such that $\|T_{y_1u_1}\|_\infty$ is minimized.

In the following we consider that V and W_n are static weighting gains and W is dynamic with the realization (A_w, B_w, C_w, D_w). Let (A_g, B_g, C_g) and (A_m, B_m, C_m) denote minimal realizations of the short period dynamics G and of the ideal model M, respectively. Then, from Figures 4.1 and 3.1 there results the following realization of the generalized system T

$$\begin{aligned} \dot{x} &= Ax + B_1u_1 + B_2u_2, \\ y_1 &= C_1x \qquad\quad + D_{12}u_2, \\ y_2 &= C_2x + D_{21}u_1, \end{aligned} \tag{4.69}$$

where $x := \begin{bmatrix} x_g^T & x_m^T & x_w^T \end{bmatrix}^T$, x_g, x_w and x_m are the state vectors of G, W, M, respectively, and

$$\begin{aligned} A &= \begin{bmatrix} A_g & 0 & 0 \\ 0 & A_m & 0 \\ -B_w C_\alpha & B_w C_m & A_w \end{bmatrix}, \\ B_1 &= \begin{bmatrix} 0 & 0 & 0 \\ B_m & 0 & 0 \\ 0 & 0 & 0 \end{bmatrix}, \qquad B_2 = \begin{bmatrix} B_g \\ 0 \\ 0 \end{bmatrix} \\ C_1 &= \begin{bmatrix} -D_w C_g & D_w C_m & C_w \\ 0 & 0 & 0 \end{bmatrix}, \quad D_{12} = \begin{bmatrix} 0 \\ V \end{bmatrix}, \\ C_2 &= \begin{bmatrix} 0 & 0 & 0 & 0 \\ C_g & 0 & 0 & 0 \end{bmatrix}, \qquad D_{21} = \begin{bmatrix} 1 & 0 \\ 0 & W_n \end{bmatrix}, \end{aligned} \tag{4.70}$$

the zero entries having appropriate dimensions and $C_\alpha := \begin{bmatrix} 1 & 0 & 0 \end{bmatrix}$.

It is easy to check that if the short period dynamics is stable, then the generalized system T satisfies the assumptions specific to the DFP (disturbance feedforward problem) discussed in the preceding chapter.

The DFP may be transformed into a two-block Nehari problem as is shown below; although this method is known (see for instance (Juang and Jonckheere, 1989)), we shall briefly describe it for completeness. In order to simplify the formulae we consider the case when $D_{21} = I$ and D_{12} is inner; notice that the more general case when D_{21} is nonsingular may be easily reduced to the case $D_{21} = I$ by rescaling the measured output y_2 by D_{21}^{-1}, and the condition D_{12} inner may be accomplished when starting from a full column rank matrix, by using the technique described in (Safonov *et al.*, 1990) based on the singular value descomposition of D_{12}.

Transforming the DFP in a two-block Nehari one (the case $D_{21} = I$ and D_{12} inner). Let X, Y be the stabilizing solutions to the Riccati equations

$$\begin{aligned} A^T X + XA - \left(XB_2 + C_1^T D_{12}\right)\left(B_2^T X + D_{12}^T C_1\right) + C_1^T C_1 &= 0 \\ AY + YA^T - YC_2^T C_2 Y + B_2 B_2^T &= 0 \end{aligned}$$

and perform the double coprime factorization

$$C_2 (sI - A_2)^{-1} B_2 = N(s)M^{-1}(s) = \widetilde{M}^{-1}(s)\widetilde{N}(s),$$

where $M, N, \widetilde{M}, \widetilde{N}, U, V, \widetilde{U}, \widetilde{V} \in \mathrm{RH}_+^\infty$ satisfy (4.52). According to (4.53), $\begin{bmatrix} M & -U \\ N & V \end{bmatrix}$ is given by

$$\left[\begin{array}{cc} M & -U \\ N & V \end{array}\right] = \left[\begin{array}{c|cc} A+B_2F & B_2 & -H \\ \hline F & I & 0 \\ C_2 & 0 & I \end{array}\right],$$

where F and H can be chosen using the stabilizing Riccati solutions X and Y, namely,

$$F := -\left(B_2^T X + D_{12}^T C_1\right)$$

and

$$H := -YC_2^T.$$

On the other hand a parameterization of all stabilizing controllers is given by (Youla *et al.*, 1976)

$$K = K_1 K_2^{-1} \tag{4.71}$$

with

$$\left[\begin{array}{c} K_1 \\ K_2 \end{array}\right] = \left[\begin{array}{cc} M & -U \\ N & V \end{array}\right] \left[\begin{array}{c} L \\ I \end{array}\right] \tag{4.72}$$

where $L \in \mathrm{RH}_+^\infty$. By coupling (4.71) to (4.69), one obtains the input–output dependence

$$T_{y_1u_1} = T_{11} + T_{12}LT_{21},$$

where

$$T_{11} = \left[\begin{array}{cc|c} A+BF_2 & -B_2F & B_1 \\ 0 & A+HC_2 & B_1+H \\ \hline C_1+D_{12}F & -D_{12}F & 0 \end{array}\right],$$

$$T_{12} = \left[\begin{array}{c|c} A+B_2F & B_2 \\ \hline C_1+D_{12}F & D_{12} \end{array}\right],$$

$$T_{21} = \left[\begin{array}{c|c} A+HC_2 & B_1+H \\ \hline C_2 & I \end{array}\right].$$

Since $A - B_1C_2$ is stable (assumption DF2), it follows that $T_{21}^{-1} \in \mathrm{RH}_+^\infty$. If we denote $\widetilde{L} := LT_{21}$ one obtains

$$T_{y_1u_1} = T_{11} + T_{12}\widetilde{L}.$$

Taking into account the expression of F it is easy to check that the conditions in Definition 3.1 are fulfilled and hence T_{12} is inner. Let $T_{12}^\perp$ be the orthogonal complement of T_{12}; then

$$\left[\begin{array}{cc} T_{12} & T_{12}^\perp \end{array}\right] = \left[\begin{array}{c|cc} A+B_2F & B_2 & -X^{-1}C_1^T D_{12}^\perp \\ \hline C_1+D_{12}F & D_{12} & D_{12}^\perp \end{array}\right],$$

where $D_{12}^{\perp}$ is the orthogonal complement of D_{12}; the invertibility condition of X required in the expression above is ensured by considering a realization of the generalized system T such that (A, B_2, C_1) is minimal, in which situation one obtains $X > 0$.

When writing $T_{y_1 u_1}$ in the equivalent form

$$T_{y_1 u_1} = T_{11} + \begin{bmatrix} T_{12} & T_{12}^{\perp} \end{bmatrix} \begin{bmatrix} \widetilde{L} \\ 0 \end{bmatrix},$$

from $\begin{bmatrix} T_{12} & T_{12}^{\perp} \end{bmatrix}$ being inner it follows that

$$\|T_{y_1 u_1}\|_{\infty} = \left\| \begin{bmatrix} T_{12} & T_{12}^{\perp} \end{bmatrix}^* T_{11} + \begin{bmatrix} \widetilde{L} \\ 0 \end{bmatrix} \right\|_{\infty}.$$

Direct calculations show that after removing the unobservable stable dynamics, one obtains

$$\begin{bmatrix} T_{12} & T_{12}^{\perp} \end{bmatrix}^* T_{11} = \left[\begin{array}{cc|c} -(A+B_2F)^T & 0 & -XB_1 \\ 0 & A+HC_2 & B_1+H \\ \hline B_2^T & -F & 0 \\ -(D_{12}^{\perp})^T C_1 X^{-1} & 0 & 0 \end{array}\right].$$

Consider now the partition

$$\begin{bmatrix} T_{12} & T_{12}^{\perp} \end{bmatrix}^* T_{11} = \begin{bmatrix} \widehat{T}_1 \\ \widehat{T}_2 \end{bmatrix},$$

where

$$\widehat{T}_1 = \left[\begin{array}{cc|c} -(A+B_2F)^T & 0 & -XB_1 \\ 0 & A+HC_2 & B_1+H \\ \hline B_2^T & -F & 0 \end{array}\right]$$

and

$$\widehat{T}_2 = \left[\begin{array}{c|c} -(A+B_2F)^T & -XB_1 \\ \hline -\left(D_{12}^{\perp}\right)^T C_1 X^{-1} & 0 \end{array}\right],$$

which is antistable. Then perform the decomposition

$$\widehat{T}_1 = \widehat{T}_{1a} + \widehat{T}_{1s}$$

with $\widehat{T}_{1a}$ antistable and $\widehat{T}_{1s}$ stable, respectively, where

$$\widehat{T}_{1a} = \left[\begin{array}{c|c} -(A+B_2F)^T & -XB_1 \\ \hline B_2^T & 0 \end{array}\right]$$

and

$$\widehat{T}_{1s} = \left[\begin{array}{c|c} A + HC_2 & B_1 + H \\ \hline -F & 0 \end{array}\right].$$

Denoting by $\widehat{L} := \widehat{T}_{1s} + \widetilde{L}$, it follows that the DFP has been reduced to a two-block Nehari problem, which consists in determining for the given antistable systems $\widehat{T}_{1a}$ and $\widehat{T}_2$ a stable system $\widehat{L}$ such that $\left\| \begin{matrix} \widehat{T}_{1a} + \widehat{L} \\ \widehat{T}_2 \end{matrix} \right\|_\infty$ is minimized. After solving this problem, we compute

$$L = \widetilde{L}T_{21}^{-1} = \left(\widehat{L} - \widehat{T}_{1s}\right) T_{21}^{-1},$$

which is replaced in (4.71) and (4.72), providing thus the optimal solution to the DFP.

Numerical results. For the model matching problem under investigation, we considered the short period dynamics of the F-4E fighter for the nominal flight conditions at altitude 35,000 ft and Mach 1.5. The aerodynamic data corresponding to the state space representation (4.67) have been derived from (Franklin and Ackermann, 1981) and they are $Z_\alpha = -0.5162\,\mathrm{rad/s}$, $Z_\delta = 0.3625\,\mathrm{rad/s}$, $M_\alpha = -18.5917\ \mathrm{rad/s^2}$, $M_q = -1.225\,\mathrm{rad/s}$, $M_\delta = -21.7304\,\mathrm{rad/s^2}$, $n_\alpha = 26.96\,\mathrm{g/rad}$, $n_\delta = -12.54\,\mathrm{g/rad}$ and the time constant $\tau = 1/14\,\mathrm{s}$, for which there results a stable short period dynamics with the eigenvalues $\{-0.87 \pm 4.3j; -14\}$.

According to the military specifications (MIL-STD-1797A, 1990) we chose for the considered nominal flight conditions an ideal model (4.68) with $\omega_m = 5\,\mathrm{rad/s}$ and $\xi_m = 0.7$. The weighting functions penalizing the tracking error and the control, have been set $W(s) = (s+3)/(s+0.03)$, $V = 1$ and $W_n = 10^{-2} \cdot I_2$, respectively.

With these numerical values we obtained a DFP corresponding to the generalized system (4.69) with the realization (4.70), which we transformed, via the procedure described above, in a two-block Nehari problem. Then we computed the optimal level of attenuation obtaining $\gamma_0 = 0.9031$ which equals for this case study, the H^∞-norm of the lower block T_2 from the Nehari problem. Hence, an optimal solution may be obtained using (4.5) corresponding to the suboptimal case since, as has been shown in Section 4.1.2, no ill-conditioned computations appear near and at the optimum if $\gamma_0 = \|T\|_2$. After computing the optimal solution to the two-block Nehari problem we obtained, using formulae (4.71) and (4.72), the optimal controller

$$K = \left[\begin{array}{c|c} A_K & B_K \\ \hline C_K & D_K \end{array}\right],$$

where

$$A_K = 10^4 \left[\; A_{K1} \quad A_{K2} \;\right],$$

with

$$A_{K1} = \begin{bmatrix} -0.0125 & 0.3545 & -0.0276 & 0.1786 \\ 0.0524 & -1.7428 & 0.1317 & -0.8781 \\ 0 & 0.0331 & -0.0020 & 0.0161 \\ 0 & 0 & 0.0002 & -0.0004 \\ 0 & 0 & 0 & 0.0002 \\ 0 & 0 & 0 & 0 \\ 0 & 0 & 0 & 0 \\ 0 & 0 & 0 & 0 \end{bmatrix},$$

$$A_{K2} = \begin{bmatrix} -0.1258 & 0.0766 & -0.4778 & 0.4225 \\ -0.6147 & -0.3539 & 0.0469 & 0.1932 \\ 0.0108 & 0.0072 & -0.1752 & 0.2106 \\ -0.0002 & 0.0018 & -0.2247 & 0.2384 \\ -0.0004 & -0.0005 & 0.0383 & -0.0231 \\ 0.0005 & -0.0012 & 0.1233 & -0.1271 \\ 0 & 0.0144 & -1.7550 & -0.0059 \\ 0 & 0 & 0.0015 & -0.0067 \end{bmatrix},$$

$$B_K = 10^3 \begin{bmatrix} -0.0001 & 0.2040 & 0 \\ 0 & -1.0019 & -0.0001 \\ -0.0005 & 0.0189 & 0 \\ 0 & 0 & 0 \\ 0.0009 & 0 & 0 \\ 0 & 0 & 0 \\ -0.0016 & 0 & 0 \\ -0.0003 & 0 & 0 \end{bmatrix},$$

$$C_K = \left[\, 0 \;\; 0 \;\; 0 \;\; 0 \;\; 0 \;\; 0 \;\; 0 \;\; 224.8543 \,\right],$$

$$D_K = \left[\, 0 \;\; 0 \;\; 0 \,\right].$$

The tracking error $e(t) = \alpha(t) - \alpha_m(t)$ and the control responses to the step reference input $\alpha_{ref} = 0.2$rad, are shown in Figures 4.2 and 4.3, respectively, indicating good steady state properties for the tracking error, achieved by an acceptable magnitude of the control.

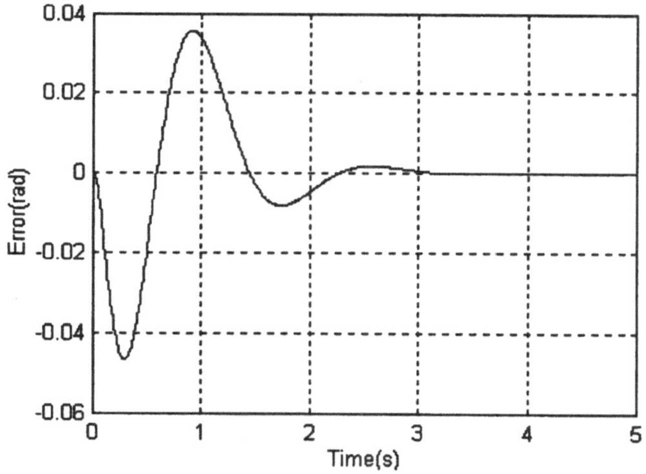

Figure 4.2. **The traking error response**

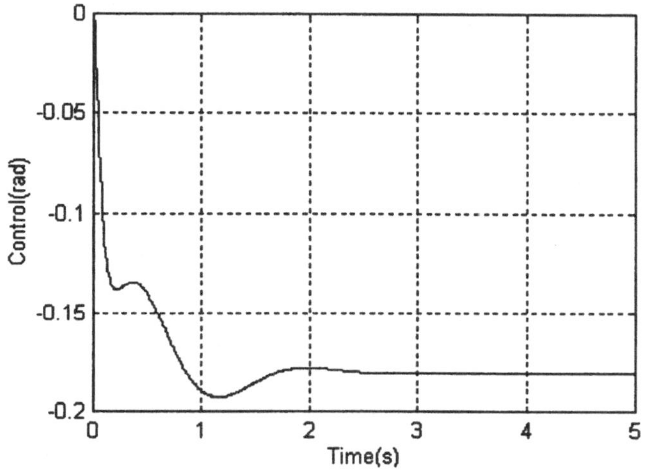

Figure 4.3. **The control response**

4.2. THE ONE-BLOCK CASE

4.2.1. SUBOPTIMAL AND OPTIMAL SOLUTIONS

In this section we consider the one-block Nehari problem and we describe one of its applications, namely the robust design with respect to normalized coprime factorization uncertainty. Given an antistable system T with the realization (A, B, C) and $\gamma > 0$, the suboptimal one-block Nehari problem consists in determining a stable system S such that

$$\|T+S\|_\infty < \gamma. \tag{4.73}$$

A solution of this problem may be directly obtained using the results from Section 4.1.1 for the two-block case, by taking $T_2 = 0$, namely $C_2 = 0$ and $D_2 = 0$. Since the solutions of equations (4.4) and (4.16) involved in solving the two-block Nehari problem reduce in the one-block case to the Lyapunov equations which give the controllability and the observability Gramians, respectively, in the following we shall denote by Q and P these solutions, which represent in fact the usual notation for the Gramians. Then, Theorem 4.1 and Corollary 4.1 directly give the following solution to the suboptimal one-block Nehari problem:

Corollary 4.2 *Let P and Q be the positive semi-definite solutions of the Lyapunov equations*

$$AP + PA^T - BB^T = 0 \tag{4.74}$$

and

$$A^TQ + QA - C^TC = 0, \tag{4.75}$$

respectively. Then, the suboptimal one-block Nehari problem has a solution if and only if

$$\rho(PQ) < \gamma^2. \tag{4.76}$$

If condition (4.76) holds then a solution of the one-block Nehari problem is given by

$$S = \left[\begin{array}{c|c} -\left(A+ZC^TC\right)^T & QB \\ \hline CZ & 0 \end{array}\right], \tag{4.77}$$

where

$$Z := P\left(\gamma^2 I - PQ\right)^{-1}. \tag{4.78}$$

□

In order to determine an optimal solution to the one-block Nehari problem, we may use the optimal solution obtained in Section 4.1.2 for the two-block case. Then, since $T_2 = 0$ in the one-block case, from Theorem 4.2 it results that the optimal Nehari distance is

$$\gamma_0 = \rho^{\frac{1}{2}}(PQ). \tag{4.79}$$

Further assume that (A, B, C) is a minimal balanced realization of T with respect to the Gramians P and Q; such a realization may be always obtained starting from an arbitrary minimal realization of T, by using the procedure described in Section 1.4. Under the balancing assumption with respect to the Gramians, we have

$$P = Q = \mathrm{diag}\,(\mu_1 I_1, ..., \mu_p I_p)$$

where $\mu_1 > ... > \mu_p > 0$ and I_k are $n_k \times n_k$ unit matrices, $k = 1, ..., p$. Notice that from the above equality and from (4.79) it results that γ_0 equals μ_1. Then the matrix Z defined by (4.78), may be written as

$$Z(\gamma) = \begin{bmatrix} Z_{11}(\gamma) & 0 \\ 0 & Z_{22}(\gamma) \end{bmatrix}, \tag{4.80}$$

where

$$Z_{11}(\gamma) = \frac{\mu_1}{\gamma^2 - \mu_1^2} I_1; \quad Z_{22}(\gamma) = \mathrm{diag}\left\{ \frac{\mu_i}{\gamma^2 - \mu_i^2} \right\}_{i=2,...,p}$$

Now, consider the following partitions of A, B and C in conformity with the structure (4.80) of $Z(\gamma)$:

$$A = \begin{bmatrix} A_{11} & A_{12} \\ A_{21} & A_{22} \end{bmatrix}; \quad B = \begin{bmatrix} B_1 \\ B_2 \end{bmatrix}; \quad C = \begin{bmatrix} C_1 & C_2 \end{bmatrix}.$$

Then an optimal solution of the one-block Nehari problem may be determined using Theorem 4.3 for the two-block case, by setting $C_{11} = C_1$; $C_{12} = C_2$, $C_{21} = 0$, $C_{22} = 0$ and $D_2 = 0$, obtaining thus:

Corollary 4.3 *Assume that $C_1^T C_1$ is nonsingular; then an optimal solution to the one-block Nehari problem is K_0 having the realization (A_0, B_0, C_0, D_0) with:*

$$\begin{aligned} A_0 &:= C_2^T C_0 - A_{22}^T, \\ B_0 &:= \gamma_0 C_2^T D_0 - R_{22} B_2, \\ C_0 &:= C_1 (C_1^T C_1)^{-1} (A_{21} + Z_{22}(\gamma_0) C_2^T C_1)^T - C_2 Z_{22}(\gamma_0), \\ D_0 &:= \gamma_0 C_1 (C_1^T C_1)^{-1} B_1, \end{aligned} \tag{4.81}$$

where $\gamma_0 = \mu_1$ and $R_{22} := \mathrm{diag}\,(\mu_2 I_2, ..., \mu_p I_p)$. □

4.2.2. OPTIMAL ROBUST STABILIZATION WITH RESPECT TO LEFT COPRIME FACTORIZATION

Given a nominal system G and $\delta > 0$, the robust stabilization problem with respect to the *normalized left coprime factorization* (NLCF) consists in determining a controller K which stabilizes all systems

$$G_\Delta = \left(\widetilde{M} + \Delta_{\widetilde{M}}\right)^{-1} \left(\tilde{N} + \Delta_{\tilde{N}}\right),$$

where $\Delta_{\widetilde{M}}, \Delta_{\widetilde{N}} \in \mathrm{RH}_+^\infty$ and $\left\| \begin{bmatrix} \Delta_{\widetilde{M}} & \Delta_{\widetilde{N}} \end{bmatrix} \right\|_\infty \leq \delta$. This problem is illustrated in Figure 4.4.

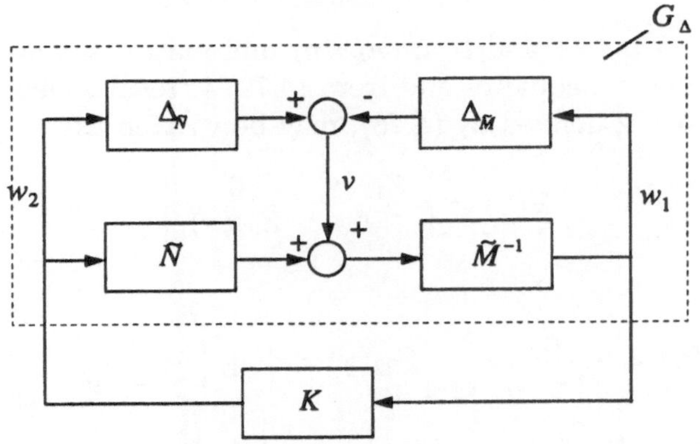

Figure 4.4. The NLCF uncertainty modeling

Let us denote by $\Delta := \begin{bmatrix} \Delta_{\widetilde{M}} & \Delta_{\widetilde{N}} \end{bmatrix} \in \mathrm{RH}_+^\infty$ and by T_{wv} the system defined by the dependence from v to $\begin{bmatrix} w_1 \\ w_2 \end{bmatrix}$ in Figure 4.4. Then it is easy to check that

$$T_{wv} = \begin{bmatrix} I \\ K \end{bmatrix} (I - GK)^{-1} \widetilde{M}^{-1}.$$

Since $\|\Delta\|_\infty \leq \delta$, from the Small Gain Theorem (Corollary 2.8) one obtains sufficient conditions for robust stability with respect to NLCF, as stated in the following proposition:

Proposition 4.1 *If K stabilizes G and*

$$\left\| \begin{bmatrix} I \\ K \end{bmatrix} (I - GK)^{-1} \widetilde{M}^{-1} \right\|_\infty < \gamma := \delta^{-1}$$

then K is a solution to the robust design problem with respect to NLCF. □

The proposition above reveals that the robust design problem with respect to NLCF may be regarded as an H^∞-control problem with the generalized system

$$T = \begin{bmatrix} T_{11} & T_{12} \\ T_{21} & T_{22} \end{bmatrix},$$

with

$$T_{11} = \begin{bmatrix} \widetilde{M}^{-1} \\ 0 \end{bmatrix}, \quad T_{12} = \begin{bmatrix} G \\ I \end{bmatrix}, \quad T_{21} = \widetilde{M}^{-1}, \quad T_{22} = G.$$

Assume for simplicity that G is strictly proper and let (A, B, C) be a realization of it. Using formulae (4.62) direct computations give

$$T = \left[\begin{array}{c|cc} A & YC^T & B \\ \hline C & I & 0 \\ 0 & 0 & I \\ \cdots & \cdots & \cdots \\ C & I & 0 \end{array}\right], \qquad (4.82)$$

where Y is the stabilizing solution of (4.57).

Remark 4.2 It can directly be checked that for the generalized system determined above the assumptions of the DFP (Section 3.6) are fulfilled and therefore the robust design problem with respect to NLCF is equivalent to a DFP formulated for (4.82). □

In the following it will be shown that the minimal level of attenuation for this DFP, denoted by γ_0, has an explicit expression in terms of the solutions X and Y to the AREs (4.56) and (4.57), respectively. According to assumption (DF3) of DFP, the KPYS(Σ_c, J_c) with Σ_c and J_c defined by (3.29) and (3.39) respectively, has a stabilizing solution (X, V_c, W_c) with $X \geq 0$; one can then consider the equivalent Popov triplet $\widetilde{\Sigma}_c$ defined by (3.40) with $\widetilde{F}_2 = -B^T X$ for which (3.41) holds, where

$$\left[\begin{array}{cc} \widetilde{T}_{11} & \widetilde{T}_{12} \end{array}\right] = \left[\begin{array}{c|cc} A - BB^T X & YC^T & B \\ \hline C & I & 0 \\ -B^T X & 0 & I \end{array}\right]. \qquad (4.83)$$

From the realization above it results, based on Definition 3.1, that $\widetilde{T}_{12}$ is inner and hence $\widetilde{T}_{12}^* \widetilde{T}_{12} = I$ which shows that the condition (3.47) holds. Hence the assumptions of Theorem 2.8 are fulfilled for $\widetilde{\Sigma}_c$ and, as has been shown in the proof of this theorem, it follows that

$$\widetilde{\mathcal{R}}_{11}^{\times} \ll 0,$$

where $\widetilde{\mathcal{R}}_{11}^{\times}$ is the Schur complement of $\widetilde{\mathcal{R}}_{22}$ in the Toeplitz operator $\widetilde{\mathcal{R}}$ associated to $\Pi_{\widetilde{\Sigma}}$, namely

$$\begin{aligned} \widetilde{\mathcal{R}} &:= \left[\begin{array}{cc} \widetilde{\mathcal{R}}_{11} & \widetilde{\mathcal{R}}_{12} \\ \widetilde{\mathcal{R}}_{12}^* & \widetilde{\mathcal{R}}_{22} \end{array}\right] \\ &= \left[\begin{array}{cc} P_+^{m_1} & 0 \\ 0 & P_+^{m_2} \end{array}\right] \\ &\times \left(\left[\begin{array}{c} \widetilde{T}_{11}^* \\ \widetilde{T}_{12}^* \end{array}\right] \left[\begin{array}{cc} \widetilde{T}_{11} & \widetilde{T}_{12} \end{array}\right] - \left[\begin{array}{cc} \gamma^2 I & 0 \\ 0 & 0 \end{array}\right] \right) \left[\begin{array}{cc} P_+^{m_1} & 0 \\ 0 & P_+^{m_2} \end{array}\right] \end{aligned}$$

$$= \left[\begin{array}{cc} P_+^{m_1}\widetilde{T}_{11}^*\widetilde{T}_{11}P_+^{m_1} - \gamma^2 P_+^{m_1} & P_+^{m_1}\widetilde{T}_{11}^*\widetilde{T}_{12}P_+^{m_2} \\ P_+^{m_2}\widetilde{T}_{12}^*\widetilde{T}_{11}P_+^{m_1} & P_+^{m_2} \end{array} \right],$$

where in our case $m_1 = p$ and $m_2 = m$, with m and p denoting the number of control inputs and output variables respectively, of the system G.

Since $\widetilde{\mathcal{R}}_{11}^{\times} \ll 0$ it results that

$$P_+^{m_1}\widetilde{T}_{11}^*\widetilde{T}_{11}P_+^{m_1} - \gamma^2 P_+^{m_1} - P_+^{m_1}\widetilde{T}_{11}^*\widetilde{T}_{12}P_+^{m_2}\widetilde{T}_{12}^*\widetilde{T}_{11}P_+^{m_1} \ll 0. \tag{4.84}$$

Further, consider the orthogonal complement $\widetilde{T}_{12}^{\perp}$ of $\widetilde{T}_{12}$ and denote by $q_1 := p_1 - m_1$. In fact for the particular DFP (4.82) $q_1 = m$. Since

$$\widetilde{T}_{12}\widetilde{T}_{12}^* + \widetilde{T}_{12}^{\perp}\left(\widetilde{T}_{12}^{\perp}\right)^* = I,$$

it results that (4.84) may be rewritten in the equivalent form

$$\begin{aligned} & P_+^{m_1}\widetilde{T}_{11}^* \left[\begin{array}{cc} \widetilde{T}_{12} & \widetilde{T}_{12}^{\perp} \end{array} \right] \left[\begin{array}{cc} P_+^{m_1} + P_-^{m_1} & 0 \\ 0 & P_+^{q_1} + P_-^{q_1} \end{array} \right] \left[\begin{array}{c} \widetilde{T}_{12}^* \\ \left(\widetilde{T}_{12}^{\perp}\right)^* \end{array} \right] \\ & \qquad \times \widetilde{T}_{11}P_+^{m_1} - \gamma^2 P_+^{m_1} - P_+^{m_1}\widetilde{T}_{11}^*\widetilde{T}_{12}P_+^{m_2}\widetilde{T}_{12}^*\widetilde{T}_{11}P_+^{m_1} \\ = \; & P_+^{m_1}\widetilde{T}_{11}^*\widetilde{T}_{12}P_-^{m_1}\widetilde{T}_{12}^*\widetilde{T}_{11}P_+^{m_1} + P_+^{m_1}\widetilde{T}_{11}^*\widetilde{T}_{12}^{\perp}\left(\widetilde{T}_{12}^{\perp}\right)^*\widetilde{T}_{11}P_+^{m_1} \\ & -\gamma^2 P_+^{m_1} = \mathbf{H}_{\widetilde{T}_{11}^*\widetilde{T}_{12}}\mathbf{H}_{\widetilde{T}_{11}^*\widetilde{T}_{12}}^* + \mathbf{T}_{\left(\widetilde{T}_{11}^*\widetilde{T}_{12}^{\perp}\right)\left(\widetilde{T}_{11}^*\widetilde{T}_{12}^{\perp}\right)^*} - \gamma^2 P_+^{m_1} \\ \ll \; & 0, \end{aligned} \tag{4.85}$$

where for the last equality we used the Definition 1.5. Hence it follows that (4.85) holds for all $\gamma > \gamma_0$ where

$$\gamma_0 = \rho^{\frac{1}{2}}\left(\mathbf{H}_{\widetilde{T}_{11}^*\widetilde{T}_{12}}\mathbf{H}_{\widetilde{T}_{11}^*\widetilde{T}_{12}}^* + \mathbf{T}_{\left(\widetilde{T}_{11}^*\widetilde{T}_{12}^{\perp}\right)\left(\widetilde{T}_{11}^*\widetilde{T}_{12}^{\perp}\right)^*}\right). \tag{4.86}$$

On the other hand, from (4.53) and (4.83) it results that

$$\left[\begin{array}{cc} \widetilde{T}_{11} & \widetilde{T}_{12} \end{array} \right] = \left[\begin{array}{cc} V & N \\ -U & M \end{array} \right],$$

from which, based on (4.51), (4.54) and (4.55), one obtains

$$\widetilde{T}_{12}^{\perp} = \left[\begin{array}{c} \widetilde{M}^* \\ -\widetilde{N}^* \end{array} \right].$$

Then

$$\widetilde{T}_{11}^*\widetilde{T}_{12}^{\perp} = V^*\widetilde{M}^* + U^*\widetilde{N}^* = I,$$

as directly follows from (4.52), and hence (4.86) becomes

$$\gamma_0 = 1 + \rho^{\frac{1}{2}}\left(\mathbf{H}_{\widetilde{T}_{11}^*\widetilde{T}_{12}}\mathbf{H}^*_{\widetilde{T}_{11}^*\widetilde{T}_{12}}\right). \tag{4.87}$$

Let us compute now $\rho^{\frac{1}{2}}\left(\mathbf{H}_{\widetilde{T}_{11}^*\widetilde{T}_{12}}\mathbf{H}^*_{\widetilde{T}_{11}^*\widetilde{T}_{12}}\right)$. To this end, using (4.83) we determine first a state space representation of $\widetilde{T}_{11}^*\widetilde{T}_{12}$, obtaining after some simple computations

$$\begin{aligned}
\dot{x}_1 &= \left(A - BB^TX\right)x_1 + Bu,\\
\dot{x}_2 &= \left(C^TC + XB^TBX\right)x_1 - \left(A - BB^TX\right)^T x_2 - XBu,\\
y &= Cx_1 - CYx_2.
\end{aligned}$$

By performing the nonsingular transformation

$$\begin{bmatrix} x_1 \\ \xi \end{bmatrix} = S\begin{bmatrix} x_1 \\ x_2 \end{bmatrix} \quad \text{with } S := \begin{bmatrix} I & 0 \\ X & I \end{bmatrix}$$

we directly obtain, using (4.56), the equivalent state space reprezentation of $\widetilde{T}_{11}^*\widetilde{T}_{12}$:

$$\begin{aligned}
\dot{x}_1 &= \left(A - BB^TX\right)x_1 \ + Bu,\\
\dot{\xi} &= -\left(A - BB^TX\right)^T \xi,\\
y &= C(I + YX)x_1 - CY\xi,
\end{aligned}$$

which reveals an uncontrollable part with the dynamics given by the second equation above; after removing this part we obtain the following state space equations of $\widetilde{T}_{11}^*\widetilde{T}_{12}$

$$\begin{aligned}
\dot{x} &= \left(A - BB^TX\right)x \ + Bu,\\
y &= C(I + YX)x.
\end{aligned} \tag{4.88}$$

According to Theorem 1.5 the Hankel norm of $\widetilde{T}_{11}^*\widetilde{T}_{12}$ is a function of its controllability and observability Gramians P and Q, respectively, defined as the unique solutions of the Lyapunov equations

$$\left(A - BB^TX\right)P + P\left(A - BB^TX\right)^T + BB^T = 0, \tag{4.89}$$

and

$$\left(A - BB^TX\right)^T Q + Q\left(A - BB^TX\right) + (I + XY)C^TC(I + YX) = 0. \tag{4.90}$$

respectively. We then have the following result:

Proposition 4.2 *The solutions of the Lyapunov equations (4.89) and (4.90) are given by*

$$P = (I + YX)^{-1} Y = Y (I + XY)^{-1} \tag{4.91}$$

and

$$Q = X (I + YX) = (I + XY) X, \tag{4.92}$$

respectively, where X and Y are the stabilizing solutions to the AREs (4.56) and (4.57), respectively.

Proof. In order to prove (4.91) consider the Popov triplet

$$\Sigma := \left(A^T, C^T; BB^T, 0, I\right)$$

for which the ARE(Σ) coincides with (4.57) and the EHP(Σ) has the form

$$\begin{bmatrix} A^T & 0 & C^T \\ -BB^T & -A & 0 \\ 0 & C & I \end{bmatrix} \begin{bmatrix} I \\ Y \\ -CY \end{bmatrix} = \begin{bmatrix} I & 0 & 0 \\ 0 & I & 0 \\ 0 & 0 & 0 \end{bmatrix} \begin{bmatrix} I \\ Y \\ -CY \end{bmatrix} \left(A - YC^TC\right)^T.$$

Pre-multiplying the above equation by

$$\begin{bmatrix} I & X & 0 \\ 0 & I & 0 \\ 0 & 0 & I \end{bmatrix}$$

it results that

$$\begin{bmatrix} A^T - XBB^T & -XA - C^TC & 0 \\ -BB^T & -A & 0 \\ 0 & C & I \end{bmatrix} \begin{bmatrix} I \\ Y \\ -CY \end{bmatrix} = \begin{bmatrix} I & X & 0 \\ 0 & I & 0 \\ 0 & 0 & 0 \end{bmatrix} \begin{bmatrix} I \\ Y \\ -CY \end{bmatrix} \left(A - YC^TC\right)^T,$$

which can be rewritten in the equivalent form

$$\begin{bmatrix} A^T - XBB^T & -XA - C^TC & 0 \\ -BB^T & -A & 0 \\ 0 & C & I \end{bmatrix} \begin{bmatrix} I & -X & 0 \\ 0 & I & 0 \\ 0 & 0 & 0 \end{bmatrix} \begin{bmatrix} I + XY \\ Y \\ -CY \end{bmatrix} = \begin{bmatrix} I & 0 & 0 \\ 0 & I & 0 \\ 0 & 0 & 0 \end{bmatrix} \begin{bmatrix} I + XY \\ Y \\ -CY \end{bmatrix} \left(A - YC^TC\right)^T.$$

Using (4.56) the equation above becomes

$$\begin{bmatrix} \left(A - BB^TX\right)^T & 0 & 0 \\ -BB^T & -\left(A - BB^TX\right) & 0 \\ 0 & C & I \end{bmatrix} \begin{bmatrix} I + XY \\ Y \\ -CY \end{bmatrix} = \begin{bmatrix} I & 0 & 0 \\ 0 & I & 0 \\ 0 & 0 & 0 \end{bmatrix} \begin{bmatrix} I + XY \\ Y \\ -CY \end{bmatrix} \left(A - YC^TC\right)^T,$$

which gives

$$\begin{bmatrix} \left(A - BB^TX\right)^T & 0 \\ -BB^T & -\left(A - BB^TX\right) \end{bmatrix} \begin{bmatrix} I + XY \\ Y \end{bmatrix} = \begin{bmatrix} I & 0 \\ 0 & I \end{bmatrix} \begin{bmatrix} I + XY \\ Y \end{bmatrix} \left(A - YC^TC\right)^T.$$

Using $I + XY > 0$, since $X \geq 0$ and $Y \geq 0$, we conclude that $Y(I + XY)^{-1}$ verifies (4.89), and then (4.91) results by uniqueness arguments.

In order to prove (4.92) one substitutes Q given by (4.92) in (4.90) and direct calculations using (4.56) give that the left hand side of (4.90) vanishes. □

Now we can state the following result:

Theorem 4.4 *The maximum stability radius for the robust design problem with respect to the NLCF is*

$$r_{\max} = \gamma_o^{-1} = [1 + \rho(XY)]^{-\frac{1}{2}}. \tag{4.93}$$

Proof. From Proposition 4.2 it results that $\rho(PQ) = \rho(XY)$ and hence (4.87) gives

$$\gamma_o = [1 + \rho(XY)]^{\frac{1}{2}},$$

from which, based on the Small Gain Theorem (Corollary 2.8)(4.93) follows immediately. □

In the following we shall determine a solution of the robust design problem with respect to NLCF for $\gamma > \gamma_0$. To this end we shall use the Remark 4.2 and the formulae (3.115) for the solution of the DFP, updated with the state space realization of T defined by (4.82). As it is shown in Theorem 3.12, the solution of the DFP depends on the stabilizing solution $(X_\gamma, V_\gamma, W_\gamma)$ of the KPYS(Σ_c, J_c) (assumption (DF3)). The following result establishes a relationship between X_γ and the stabilizing solutions X and Y to (4.56) and (4.57), respectively.

Proposition 4.3 *Let $\gamma > [1+\rho(XY)]^{\frac{1}{2}}$. Then*

$$X_\gamma = -\gamma^2 X\left[\left(1-\gamma^2\right)I + YX\right]^{-1}. \tag{4.94}$$

Proof. The Popov triplet Σ_c defined by (3.29) updated with (4.82) becomes

$$\Sigma_c := \left(A, \begin{bmatrix} YC^T & B \end{bmatrix}; C^TC, \begin{bmatrix} C^T & 0 \end{bmatrix}, \begin{bmatrix} -\left(\gamma^2-1\right)I_p & 0 \\ 0 & I_m \end{bmatrix}\right),$$

and the KPYS(Σ_c, J_c) with

$$J_c = \begin{bmatrix} -I_p & 0 \\ 0 & I_m \end{bmatrix},$$

has, according to (DF3), a stabilizing solution $(X_\gamma, V_\gamma, W_\gamma)$. From Theorem 2.7 it results that this is equivalent to the EHP(Σ_c) being regular and stable disconjugate, namely the following equality holds

$$N_cV = M_cVS, \tag{4.95}$$

where

$$\begin{aligned} M_c &= \begin{bmatrix} I & 0 & 0 & 0 \\ 0 & I & 0 & 0 \\ 0 & 0 & 0 & 0 \\ 0 & 0 & 0 & 0 \end{bmatrix}, \\ N_c &= \begin{bmatrix} A & 0 & YC^T & B \\ -C^TC & -A^T & -C^T & 0 \\ C & CY & -\left(\gamma^2-1\right)I & 0 \\ 0 & B^T & 0 & I \end{bmatrix}, \\ V &= \begin{bmatrix} I \\ X_\gamma \\ F_1 \\ F_2 \end{bmatrix}, \end{aligned} \tag{4.96}$$

and $S := A + YC^TF_1 + BF_2$ is stable.

Let us define the unimodular matrices

$$U = \begin{bmatrix} U_1 & U_2 \end{bmatrix},$$

where

$$U_1 := \begin{bmatrix} \left(1-\gamma^{-2}\right)^{-1}I & \gamma^{-2}\left(1-\gamma^{-2}\right)^{-1}Y \\ 0 & I \\ 0 & 0 \\ 0 & 0 \end{bmatrix},$$

$$U_2 \quad := \quad \begin{bmatrix} -\left(1-\gamma^2\right)^{-1} YC^T & -\gamma^2\left(1-\gamma^{-2}\right)^{-1} B \\ \left(1-\gamma^2\right)^{-1} C^T & 0 \\ 0 & I \\ \left(1-\gamma^2\right)^{-1} I & 0 \end{bmatrix},$$

and

$$W = \begin{bmatrix} \left(1-\gamma^{-2}\right) I & -\gamma^{-2} Y & 0 & 0 \\ 0 & I & 0 & 0 \\ \gamma^{-2} C & \gamma^{-2} CY & 0 & I \\ 0 & 0 & I & 0 \end{bmatrix}.$$

When pre-multiplying (4.95) by U one obtains

$$\widetilde{N}_c \widetilde{V} = \widetilde{M}_c \widetilde{V} S, \tag{4.97}$$

where

$$\begin{aligned} \widetilde{M}_c \quad &:= \quad UM_cW = \begin{bmatrix} I & 0 & 0 & 0 \\ 0 & I & 0 & 0 \\ 0 & 0 & 0 & 0 \\ 0 & 0 & 0 & 0 \end{bmatrix}, \\ \widetilde{N}_c \quad &:= \quad UN_cW = \begin{bmatrix} A & 0 & B & 0 \\ -C^TC & -A^T & 0 & 0 \\ 0 & B^T & I & 0 \\ 0 & 0 & 0 & I \end{bmatrix}, \\ \widetilde{V} \quad &:= \quad W^{-1}V = \begin{bmatrix} \left(\gamma^2-1\right)^{-1}\left(\gamma^2 I + YX_\gamma\right) \\ X_\gamma \\ F_2 \\ * \end{bmatrix}. \end{aligned}$$

The first 3×3 block entries in (4.97) give

$$\begin{bmatrix} A & 0 & B \\ -C^TC & -A^T & 0 \\ 0 & B^T & I \end{bmatrix} \begin{bmatrix} I \\ \widetilde{X}_\gamma \\ \widetilde{F}_2 \end{bmatrix} = \begin{bmatrix} I & 0 & 0 \\ 0 & I & 0 \\ 0 & 0 & 0 \end{bmatrix} \begin{bmatrix} I \\ \widetilde{X}_\gamma \\ \widetilde{F}_2 \end{bmatrix} \widetilde{S} \tag{4.98}$$

where

$$\begin{aligned} \widetilde{X}_\gamma \quad &:= \quad \left(\gamma^2-1\right) X_\gamma \left(\gamma^2 I + YX_\gamma\right)^{-1}, \\ \widetilde{F}_2 \quad &:= \quad \left(\gamma^2-1\right) F_2 \left(\gamma^2 I + YX_\gamma\right)^{-1}, \\ \widetilde{S} \quad &:= \quad \left(\gamma^2 I + YX_\gamma\right) S \left(\gamma^2 I + YX_\gamma\right)^{-1}. \end{aligned} \tag{4.99}$$

Since (4.98) is just the EHP associated with (4.56), by a uniqueness argument it follows that

$$X = \widetilde{X}_\gamma, \quad \widetilde{F}_2 = -B^T X, \quad \widetilde{S} = A - BB^T X.$$

Using (4.99) it results that

$$X = \left(\gamma^2 - 1\right) X_\gamma \left(\gamma^2 I + Y X_\gamma\right)^{-1},$$

from which (4.94) immediately results. □

Now we can state the following result which provides a solution to the robust design problem with respect to NLCF.

Theorem 4.5 *The controller K having the realization (A_K, B_K, C_K) with*

$$\begin{aligned} A_K &= A - YC^TC + \gamma^2 BB^T X \left[\left(1-\gamma^2\right) I + YX\right]^{-1}, \\ B_K &= YC^T, \\ C_K &= \gamma^2 B^T X \left[\left(1-\gamma^2\right) I + YX\right]^{-1}. \end{aligned} \tag{4.100}$$

is a solution of the robust design problem with respect to NLCF for $\gamma > [1+\rho(XY)]^{\frac{1}{2}}$.

Proof. Since according to Remark 4.2, the generalized system (4.82) satisfies the assumptions about the DFP, it follows that a solution of the robust design with respect to NLCF may be determined using formulae (3.115) updated with (4.82); these formulae directly give (4.100). □

The following result emphasizes a relationship between the robust design problem with respect to NLCF and the one-block Nehari problem.

Theorem 4.6 *Let $\gamma > \gamma_0 = [1+\rho(XY)]^{\frac{1}{2}}$; then a solution of the robust design problem with respect to NLCF is given by*

$$K = N_K M_K^{-1}, \tag{4.101}$$

where $\begin{bmatrix} N_K \\ M_K \end{bmatrix}$ is the solution to the one-block Nehari problem

$$\left\| \begin{bmatrix} -\widetilde{N}^* \\ \widetilde{M}^* \end{bmatrix} + \begin{bmatrix} N_K \\ M_K \end{bmatrix} \right\|_\infty \le \left(1-\gamma^{-2}\right)^{\frac{1}{2}}. \tag{4.102}$$

Proof. From (4.62) it results that

$$\widetilde{M}^* = \left[\begin{array}{c|c} -\left(A-YC^TC\right)^T & C^T \\ \hline CY & I \end{array}\right],$$

$$\widetilde{N}^* = \left[\begin{array}{c|c} -\left(A-YC^TC\right)^T & -C^T \\ \hline B^T & 0 \end{array}\right].$$

Then based on Corollary 4.2, a solution to (4.102) can be determined using (4.77) updated with

$$\left[\begin{array}{c} -\widetilde{N}^* \\ \widetilde{M}^* \end{array}\right] = \left[\begin{array}{c|c} -\left(A-YC^TC\right)^T & C^T \\ \hline B^T & 0 \\ CY & I \end{array}\right]. \tag{4.103}$$

First, let us determine the controllability and observability Gramians $\widetilde{P}$ and $\widetilde{Q}$ respectively of (4.103). Using (4.56) and (4.57), it is easy to check that

$$\widetilde{P} = X(I+YX)^{-1}, \quad \widetilde{Q} = Y, \tag{4.104}$$

and hence

$$\begin{aligned} \rho\left(\widetilde{P}\widetilde{Q}\right) &= \lambda_{\max}\left(YX\left(I+YX\right)^{-1}\right) \\ &= 1-\left[1-\lambda_{\max}\left(YX\left(I+YX\right)^{-1}\right)\right] \\ &= 1-\left[1+\lambda_{\min}\left(-YX\left(I+YX\right)^{-1}\right)\right] \\ &= 1-\lambda_{\min}\left(I-YX\left(I+YX\right)^{-1}\right) \\ &= 1-\lambda_{\min}\left(I+YX\right)^{-1} = 1-\frac{1}{\lambda_{\max}(I+YX)} \\ &= 1-\frac{1}{1+\rho\left(XY\right)} = 1-\gamma_0^{-2}. \end{aligned} \tag{4.105}$$

According to Corollary 4.2 the one-block Nehari problem (4.102) has a solution if and only if $\rho^{\frac{1}{2}}\left(\widetilde{P}\widetilde{Q}\right) < \tilde{\gamma} := (1-\gamma^{-2})^{\frac{1}{2}}$, and hence from (4.105) it results that the problem can be solved only for $\tilde{\gamma} > \left(1-\gamma_0^{-2}\right)^{\frac{1}{2}}$. The solution (4.77) to the one-block Nehari problem is expressed as function of Z defined by (4.78) which in our case, becomes together with (4.104):

$$\begin{aligned}
\widetilde{Z} &= \widetilde{P}\left(\tilde{\gamma}^2 I - \widetilde{P}\widetilde{Q}\right)^{-1} \\
&= X(I+YX)^{-1}\left[\left(1-\gamma^{-2}\right)I - YX(I+YX)^{-1}\right]^{-1} \\
&= \left(1-\gamma^{-2}\right)^{-1} X(I+YX)^{-1} \\
&\quad \times \left[I - \left(1-\gamma^{-2}\right)^{-1} YX(I+YX)^{-1}\right]^{-1} \\
&= X\left[\left(1-\gamma^{-2}\right)(I+YX) - YX\right]^{-1} \\
&= -\gamma^2 X\left[\left(1-\gamma^2\right)I + YX\right]^{-1} = X_\gamma
\end{aligned}$$

where for the last equality above we have used (4.94). Then (4.77) applied to (4.103) and $\widetilde{Z} = X_\gamma$ gives the following solution of (4.102)

$$\begin{bmatrix} N_K \\ M_K \end{bmatrix} = \left[\begin{array}{c|c} A - YC^TC - \left(BB^T + YC^TCY\right)X_\gamma & YC^T \\ \hline \begin{bmatrix} B^T \\ CY \end{bmatrix} X_\gamma & \begin{matrix} 0 \\ -I \end{matrix} \end{array}\right],$$

where $A - YC^TC - \left(BB^T + YC^TCY\right)X_\gamma$ is stable. From the expression above for M_K it results that

$$M_K^{-1} = \left[\begin{array}{c|c} A - YC^TC - BB^TX_\gamma & -YC^T \\ \hline CYX_\gamma & -I \end{array}\right],$$

and therefore one obtains the following state space representation of $N_K M_K^{-1}$:

$$\begin{aligned}
\dot{x}_1 &= \left(A - YC^TC - BB^TX_\gamma\right)x_1 - YC^Tu, \\
\dot{x}_2 &= YC^TCYX_\gamma x_1 + \left[A - YC^TC - \left(BB^T + YC^TCY\right)X_\gamma\right]x_2 \\
&\quad -YC^Tu, \\
y &= B^TX_\gamma x_2,
\end{aligned}$$

from which, by performing the nonsingular state transformation

$$\begin{bmatrix} x_1 \\ \xi \end{bmatrix} = \begin{bmatrix} I & 0 \\ I & -I \end{bmatrix}\begin{bmatrix} x_1 \\ x_2 \end{bmatrix},$$

there results the equivalent representation

$$\begin{aligned}
\dot{x}_1 &= \left(A - YC^TC - BB^TX_\gamma\right)x_1 - YC^Tu, \\
\dot{\xi} &= \left[A - YC^TC - \left(BB^T + YC^TCY\right)X_\gamma\right]\xi, \\
y &= B^TX_\gamma(x_1 - \xi),
\end{aligned}$$

which reveals an uncontrollable stable dynamics; after removing it we obtain

$$N_K M_K^{-1} = \left[\begin{array}{c|c} A - YC^TC - BB^TX_\gamma & -YC^T \\ \hline B^TX_\gamma & 0 \end{array}\right].$$

Owing to (4.94), the expression above for $N_K M_K^{-1}$ coincides with that of K given by (4.100), which represents the solution of the robust design problem with respect to NLCF. □

Consider now the optimal solution to the Nehari problem (4.102) which has the form

$$\left[\begin{array}{c} N_{K_0} \\ M_{K_0} \end{array}\right] = \left[\begin{array}{c|c} A_0 & B_0 \\ \hline C_{01} & D_{01} \\ C_{02} & D_{02} \end{array}\right],$$

where A_0 is stable. Since a realization of $M_{K_0}^{-1}$ is

$$\left(A_0 - B_0 D_{02}^{-1} C_{02}, B_0 D_{02}^{-1}, -D_{02}^{-1} C_{02}, D_{02}^{-1}\right),$$

the optimal controller $K_0 = N_{K_0} M_{K_0}^{-1}$ has the following state space equations

$$\begin{aligned} \dot{x}_1 &= \left(A_0 - B_0 D_{02}^{-1} C_{02}\right) x_1 + B_0 D_{02}^{-1} u, \\ \dot{x}_2 &= -B_0 D_{02}^{-1} C_0 x_1 + A_0 x_2 + B_0 D_{02}^{-1} u, \\ y &= -D_{01} D_{02}^{-1} C_{02} x_1 + C_{01} x_2 + D_{01} D_{02}^{-1} u. \end{aligned} \tag{4.106}$$

By subtracting the first two equations above, one obtains

$$\dot{x}_1 - \dot{x}_2 = A_0(x_1 - x_2),$$

which shows, taking into account that A_0 is stable, that the system (4.106) has a stable uncontrollable part; after the removal of this part we obtain the following optimal robust controller

$$K_0 = \left[\begin{array}{c|c} A_0 - B_0 D_{02}^{-1} C_{02} & B_0 D_{02}^{-1} \\ \hline C_{01} - D_{01} D_{02}^{-1} C_{02} & D_{01} D_{02}^{-1} \end{array}\right]. \tag{4.107}$$

The realization above may be expressed in an explicit form with respect to the matrices A, B and C if these matrices give a balanced realization of the nominal system G with respect to the solutions of the standard AREs (4.56) and (4.57). To this end, recall first that the eigenvalues of the matrix XY are invariant with respect to nonsingular coordinate transformations, as can be easily proved, and they are called *Jonckheere–Silverman invariants*. A minimal realization (A, B, C) is said *balanced with respect to X and* Y (or *in the sense of Jonckheere–Silverman*) if the standard AREs (4.56) and (4.57) have the solutions $X = Y = \Sigma = \operatorname{diag}(\sigma_1 I_1, ..., \sigma_r I_r)$ respectively, with $\sigma_1 > ... > \sigma_r > 0$, where I_k are $n_k \times n_k$ unit matrices, $k = 1, ..., r$. Notice that for any arbitrary minimal realization (A, B, C) of G, an equivalent balanced realization in the sense of Jonckheere–Silverman may be easily obtained via the transformation $\widehat{T} := \Sigma^{-\frac{1}{2}} R^T \widehat{Z}$ where R, Σ are given by the singular value descomposition $\widehat{Z} Y \widehat{Z}^T = R \Sigma^2 R^T$ with R orthogonal, and $\widehat{Z}$ is provided by the Cholesky factorization $X = \widehat{Z}^T \widehat{Z}$. Therefore, without loss of generality, we may consider that (A, B, C) is a balanced realization of G with respect to X and Y.

In order to determine the solution of the optimal one-block Nehari problem, we shall use the result given by Corollary 4.3 in which a balanced realization of the system (4.103) in the sense of Moore, is required. Taking into account that $X = Y = \Sigma$ and using (4.56) and (4.57), direct calculations show that the controllability and the observability Gramians of $\begin{bmatrix} -\widetilde{N}^* \\ \widetilde{M}^* \end{bmatrix}$ are $P = \Sigma \left(I + \Sigma^2\right)^{-1}$ and $Q = \Sigma$, respectively. Then the transformation $T_b := \left(I + \Sigma^2\right)^{\frac{1}{4}}$ applied to the system (4.103) gives a balanced realization of $\begin{bmatrix} -\widetilde{N}^* \\ \widetilde{M}^* \end{bmatrix}$ in the sense of Moore, obtaining thus the Gramians

$$P_b = T_b^{-T} P T_b^{-1} = Q_b = T_b Q T_b^T = \Sigma \left(I + \Sigma^2\right)^{-\frac{1}{2}} .$$

Assuming that $C_1^T C_1$ is nonsingular, formulae (4.81) give the optimal solution to the one-block Nehari problem (4.102), $\begin{bmatrix} N_{K_0} \\ M_{K_0} \end{bmatrix} = \left[\begin{array}{c|c} A_0 & B_0 \\ \hline C_0 & D_0 \end{array}\right]$ with

$$\begin{aligned}
A_0 &= \left(I + \Sigma_{22}^2\right)^{-\frac{1}{4}} \left\{ \left(A_{22} - \Sigma_{22} C_2^T C_2\right) \left(I + \Sigma_{22}^2\right)^{\frac{1}{4}} \right. \\
&\qquad \left. - \begin{bmatrix} B_2 & \Sigma_{22} C_2^T \end{bmatrix} C_0 \right\}, \\
B_0 &= -\sigma_1 (1 + \sigma_1^2)^{-1} \left(I + \Sigma_{22}^2\right)^{-\frac{1}{4}} \left(B_2 B_1^T + \sigma_1 \Sigma_{22} C_2^T C_1\right) \\
&\qquad \times \left(C_1^T C_1\right)^{-1} C_1^T + R_{22} \left(I + \Sigma_{22}^2\right)^{\frac{1}{4}} C_2^T, \\
C_0 &= \begin{bmatrix} C_{01} \\ C_{02} \end{bmatrix} \qquad\qquad (4.108) \\
&= (1 + \sigma_1^2)^{-1} \begin{bmatrix} B_1^T \\ \sigma_1 C_1 \end{bmatrix} \left(C_1^T C_1\right)^{-1} \left[\left(A_{12} - \sigma_1 C_1^T C_2\right) \right. \\
&\qquad \left. \times \left(I + \Sigma_{22}^2\right)^{\frac{1}{4}} - \left(B_1 B_2^T + \sigma_1 C_1^T C_2 \Sigma_{22}\right) \left(I + \Sigma_{22}^2\right)^{-\frac{1}{4}} Z_{22} \right]
\end{aligned}$$

$$
\begin{aligned}
& + \left[\begin{array}{c} B_2^T \\ C_2 \Sigma_{22} \end{array} \right] \left(I + \Sigma_{22}^2 \right)^{-\frac{1}{4}} Z_{22}, \\
D_0 &= \left[\begin{array}{c} D_{01} \\ D_{02} \end{array} \right] = \sigma_1 (1 + \sigma_1^2)^{-1} \left[\begin{array}{c} B_1^T \\ \sigma_1 C_1 \end{array} \right] \left(C_1^T C_1 \right)^{-1} C_1^T - \left[\begin{array}{c} 0 \\ I \end{array} \right],
\end{aligned}
$$

where

$$
\begin{aligned}
Z_{22} &:= (1 + \sigma_1^2) \operatorname{diag} \left\{ \frac{\sigma_i \sqrt{1 + \sigma_i^2}}{\sigma_1^2 - \sigma_i^2} \right\}_{i=2,\dots,p}, \\
\Sigma_{22} &:= \operatorname{diag} \{ \sigma_2 I_2, \dots, \sigma_p I_p \}_{i=2,\dots,p}, \\
R_{22} &:= \operatorname{diag} \left\{ \frac{\sigma_i}{\sqrt{1 + \sigma_i^2}} \right\}_{i=2,\dots,p}.
\end{aligned}
$$

Therefore, the algorithm to compute an optimal robust controller with respect to NLCF can be summarized as follows:

Step 1 Determine a minimal balanced realization (A, B, C) of the nominal plant in the sense of Jonckheere–Silverman and perform the partition of A, B, C conformably to the structure of $X = Y = \Sigma = \operatorname{diag}(\sigma_1 I_1, \dots \sigma_p I_p)$;

Step 2 Compute an optimal solution to the one-block Nehari problem (4.102) using formulae (4.108);

Step 3 Determine the optimal robust controller K_0 by (4.107).

4.2.3. A NUMERICAL EXAMPLE

We consider a *loop shaping* problem solved using the approach proposed in (McFarlane and Glover, 1990); this method consists in determining first a 'shaped' system $G_s = W_2 G W_1$, where G denotes the nominal plant and W_1 and W_2 are shaping functions chosen such that the magnitude diagrams of the singular values of G_s satisfy certain requirements imposed by sensitivity and robustness design objectives. Further, a robust controller with respect to NLCF, denoted by K_∞, is computed and finally, the solution of the loop shaping problem is given by $K = W_1 K_\infty W_2$ (details are given in (McFarlane and Glover, 1990)).

In our case study, an optimal robust controller will be determined using the method described in Section 4.2.2. The nominal system G is the linearized longitudinal dynamics of an aircraft, also considered in (McFarlane and Glover, 1990), with the state space realization (A, B, C, D), where

$$A = \begin{bmatrix} 0 & 0 & 1.1320 & 0 & -1.0000 \\ 0 & -0.0538 & -0.1712 & 0 & 0.0705 \\ 0 & 0 & 0 & 1.0000 & 0 \\ 0 & 0.0485 & 0 & -0.8556 & -1.0130 \\ 0 & -0.2909 & 0 & 1.0532 & -0.6859 \end{bmatrix},$$

$$B = \begin{bmatrix} 0 & 0 & 0 \\ -0.1200 & 1.0000 & 0 \\ 0 & 0 & 0 \\ 4.4190 & 0 & -1.6650 \\ 1.5750 & 0 & -0.0732 \end{bmatrix},$$

$$C = \begin{bmatrix} 1 & 0 & 0 & 0 & 0 \\ 0 & 1 & 0 & 0 & 0 \\ 0 & 0 & 1 & 0 & 0 \end{bmatrix},$$

$$D = \begin{bmatrix} 0 & 0 & 0 \\ 0 & 0 & 0 \\ 0 & 0 & 0 \end{bmatrix}.$$

The state variables are: x_1 relative altitude (m), x_2 forward speed (m/s), x_3 pitch angle (deg), x_4 pitch rate (deg /s), x_5 vertical speed (m/s), the control inputs are u_1 spoiler angle (10^{-1} deg), u_2 forward acceleration (m/s^2), u_3 elevator angle (deg) and the measured outputs are the first three states. In the design example considered, the shaping functions

$$W_1 = I, \quad W_2 = \mathrm{diag}\left\{24\frac{s+0.4}{s}, 12\frac{s+0.4}{s}, 24\frac{s+0.4}{s}\right\}$$

has been used.

We have determined the 'shaped' system G_s for which we computed the optimal stability radius with respect to left coprime factorization $r_{\max} = \gamma_0^{-1} = (1+\sigma_1^2)^{-\frac{1}{2}} = 0.3786$. Using the algorithm described in Section 4.2.2, we obtained by (4.108) the seventh-order optimal robust controller K_∞ having the realization $(A_{K_\infty}, B_{K_\infty}, C_{K_\infty}, D_{K_\infty})$, where

$$A_{K_\infty} = \begin{bmatrix} A_{K_{\infty 1}} & A_{K_{\infty 2}} \end{bmatrix},$$

$$A_{K_{\infty 1}} = \begin{bmatrix} -114.9672 & 0.8004 & -2.3752 & 9.2718 \\ -12.9132 & -29.5942 & 0.5337 & -0.1790 \\ 41.5348 & -0.5943 & -7.3158 & -1.9989 \\ -163.7037 & -1.3762 & -1.6163 & -12.0343 \\ 4.6008 & 1.3511 & 0.1793 & 0.5973 \\ -4.9282 & 0.3554 & -0.1748 & -0.6349 \\ 1.3144 & -0.0190 & -0.2475 & -0.0205 \end{bmatrix},$$

$$
\begin{aligned}
A_{K_{\infty 2}} &= \begin{bmatrix} 0.1512 & -0.1828 & -0.1063 \\ -0.3776 & -0.4071 & 0.0719 \\ 0.9057 & -0.1984 & 0.0654 \\ 0.2450 & 0.1788 & 0.2433 \\ -0.4129 & 0.0141 & -0.0152 \\ 0.0304 & -0.3880 & 0.0118 \\ 0.0318 & -0.0078 & -0.3985 \end{bmatrix}, \\
B_{K_\infty} &= \begin{bmatrix} 1.8393 & 0.1341 & -2.7303 \\ -0.3115 & 2.9677 & -0.2042 \\ 2.8197 & 0.1027 & 1.5882 \\ 1.1956 & 0.0981 & 1.7941 \\ -0.0800 & -0.0017 & -0.0512 \\ 0.0855 & 0.0043 & 0.0557 \\ 0.0622 & 0.0026 & 0.0408 \end{bmatrix}, \\
C_{K_\infty} &= \begin{bmatrix} C_{K_{\infty 1}} & C_{K_{\infty 2}} \end{bmatrix}, \\
C_{K_{\infty 1}} &= \begin{bmatrix} 74.6135 & 0.5069 & -0.5969 & 0.8948 \\ 2.1856 & -6.0204 & 0.0751 & 0.1037 \\ -4.5229 & -0.2643 & -4.1829 & -2.6444 \end{bmatrix}, \\
C_{K_{\infty 2}} &= \begin{bmatrix} 0.0258 & -0.0039 & 0.0035 \\ -0.0083 & 0.0025 & -0.0016 \\ 0.3749 & -0.0619 & 0.0476 \end{bmatrix}, \\
D_{K_\infty} &= \begin{bmatrix} 0.1514 & 0.0064 & 0.0994 \\ -0.0677 & -0.0028 & -0.0444 \\ 2.0354 & 0.0857 & 1.3366 \end{bmatrix}.
\end{aligned}
$$

As has been shown in the previous section, the robust design problem with respect to NLCF is equivalent, in the case $D = 0$ with the H^∞-control problem where T defined is by (4.82). This fact provides a method of evaluating the robustness performances with respect to NLCF, achieved with a stabilizing controller K; to this end one computes the H^∞-norm of the resulting system obtained by coupling K to (4.82) and the inverse of this norm is just the robustness radius. We performed such an evaluation for the controller K_∞ given above, by using the generalized system (4.82) updated to G_s and we obtained a resulting system with the poles:

$$
\begin{gathered}
-99.7302 \\
-17.2174 \\
-7.6841 \pm 7.7107j \\
-2.3730 \pm 2.3478j \\
-11.8993 \\
-3.3436
\end{gathered}
$$

$$\begin{array}{l} -12.0509 \\ -0.3962 \\ -0.3995 \pm 0.0023j \\ -0.3994 \\ -0.4001 \\ -0.4000 \end{array}$$

which shows that K_∞ is indeed stabilizing. The H^∞-norm of the resulting system equals γ_0. Further, the solution to the loop shaping problem is given by $K = W_2 K_\infty$.

NOTES AND REFERENCES

The Nehari problem originated in (Nehari, 1957) and an historical account of the development of its solutions can be found, for instance, in (Safonov *et al.*, 1987). The wide variety of methods of solving this problem includes interpolation techniques (Ball *et al.*, 1990), (Gohberg *et al.*, 1991), (Dym and Gohberg, 1988), γ-procedures (Francis, 1983), (Doyle, 1983), (Chu *et al.*, 1986) and approaches based on the generalized Popov–Yakubovich theory (Ionescu and Oara, 1996b).

State space suboptimal solutions are given in (Jonckheere *et al.*, 1989), (Halanay, 1990), (Dragan *et al.*, 1995a), (Ionescu and Oara, 1996b).

Different methods have been proposed for computing the optimal Nehari distance; we mention among them those based on the γ-iteration approach (Doyle, 1983), (Chang and Pearson, 1985) and those using Hankel and Toeplitz operators (Glover, 1984), (Feintuch and Francis, 1985), (Jonckheere and Juang, 1987).

In (Glover *et al.*, 1991) and (Safonov *et al.*, 1990) optimal solutions of the Nehari problem in a descriptor form are obtained. Alternative methods for the optimal solution are proposed in (Dragan *et al.*, 1995a,b,c) where a singular perturbations based technique is used. The optimal solutions derived in Sections 4.1.2 and 4.2.1 are essentially based on this method.

More details concerning the coprime factorization and the uncertainty modeling based on it, can be found, for instance, in (Nett *et al.*, 1984), (Francis, 1986), (Vidyasagar and Kimura, 1986) and (McFarlane and Glover, 1990) and in the their references. The result in Theorem 4.4 has been proved by a different technique in (McFarlane and Glover, 1990). The method used in our book is similar to that described in (Ionescu, 1995) for the discrete time case.

A standard reference for the Jonckheere-Silverman invariants used in Section 4.2.2 is (Jonckheere and Silverman, 1983). Similar formulae to (4.108) are derived in (Dragan *et al.*, 1995b).

CHAPTER 5

OPTIMAL H^∞ PROBLEMS: A SINGULAR PERTURBATION APPROACH

The results derived in Chapter 3 allow us to determine a suboptimal solution of the H^∞-control problem which depends on the imposed level of attenuation γ. In order to improve the attenuating performances of the resulting system, it is desirable to design the H^∞ controllers for low values of γ, as close as possible to their minimum γ_0. The optimal level of attenuation γ_0 is in fact the largest γ for which one of the necessary and sufficient conditions for the solvability of the H^∞-control problem given by Theorems 3.7 and 3.9 fails; therefore the following cases may occur at γ_0:

(C1) The KPYS(Σ_c, J_c) or KPYS(Σ_o, J_o) has no more stabilizing solution; or

(C2) The spectral radius condition fails, namely γ_0 satisfies the equation

$$\rho\left(X(\gamma_0)Y(\gamma_0)\right) = \gamma_0^2. \tag{5.1}$$

In the first case above one can directly use the explicit state space formulae for the suboptimal H^∞ controller for any $\gamma > \gamma_0$, no matter how close γ is to γ_0.

When the optimum γ_0 verifies (5.1) the state space formulae for the H^∞ controller give ill-conditioned computations for γ close to γ_0, since $\gamma^2 I - X(\gamma)Y(\gamma)$ is singular at γ_0 and therefore the matrix $\left(\gamma^2 I - X(\gamma)Y(\gamma)\right)^{-1}$ entering in the realization of the H^∞ controller tends to become unbounded when γ tends to γ_0.

In order to avoid this unpleasant behavior near the optimum, different methods have been proposed, a brief discussion of them being presented in the section Notes and References. In this chapter we shall describe an alternative approach based the theory of singularly perturbed systems, which allows us to determine an optimal solution of the H^∞-control problem.

5.1. OPTIMAL SOLUTION OF THE H^∞-CONTROL PROBLEM

For reasons of simplicity of formulae, first we consider the so called *standard H^∞*-control problem for which the normalizing conditions (3.52) hold.

Then, based on the developments in Chapter 3, the following assumptions are considered in this case:

(A1) The pairs (A, B_2) and (C_2, A) are stabilizable and detectable, respectively;

(A2) The normalizing conditions (3.52) are satisfied;

(A3) (A, B_1, C_1) is minimal.

Remark 5.1 Assumption (A3) is stronger than the regularity conditions (R1) and (R2) imposed in Section 3.1, and it is considered just for convenience; we shall show further the modifications required to relax this assumption to (R1) and (R2). □

From Theorems 3.8, 3.9 and 3.10 it results that the standard H^∞-control problem corresponding to the unscaled case ($\gamma \neq 1$) has a solution if and only if, the AREs

$$A^T X + XA + X\left(\gamma^{-2}B_1B_1^T - B_2B_2^T\right)X + C_1^T C_1 = 0, \qquad (5.2)$$
$$AY + YA^T + Y\left(\gamma^{-2}C_1^T C_1 - C_2^T C_2\right)Y + B_1 B_1^T = 0 \qquad (5.3)$$

have stabilizing positive semi-definite solutions, and

$$\rho(XY) < \gamma^2. \qquad (5.4)$$

If these conditions are satisfied, then according to Remark 3.2, a sub-optimal solution to the H^∞-control problem is given by

$$K = \left[\begin{array}{c|c} A_K & B_K \\ \hline C_K & 0 \end{array}\right],$$

where

$$\begin{aligned} A_K &= A + \left(\gamma^{-2}B_1B_1^T - B_2B_2^T\right)X - ZC_2^T C_2, \\ B_K &= ZC_2^T, \\ C_K &= -B_2^T X; \quad Z := \gamma^2 Y\left(\gamma^2 I - XY\right)^{-1}. \end{aligned} \qquad (5.5)$$

A known result concerning the dependence of the stabilizing solutions X and Y with respect to γ, allowing us to compute an optimal level of attenuation γ_0, is the following lemma:

Lemma 5.1 *The functions $\gamma \mapsto X(\gamma)$ and $\gamma \mapsto Y(\gamma)$ are decreasing.*

Proof. Let $X(\gamma)$ be the stabilizing solution of (5.2) and consider the KPYS(Σ_c, J_c) with J_c given by (3.39) and $\Sigma_c := (A, B; Q_c, L_c, R_c)$ defined by (3.29), where Q_c, L_c and R_c become under the normalizing conditions (3.52) for $\gamma \neq 1$,

$$\begin{aligned} Q_c &= C_1^T C_1, \\ L_c &= \begin{bmatrix} 0 & 0 \end{bmatrix}, \\ R_c &= \begin{bmatrix} -\gamma^2 I_{m_1} & 0 \\ 0 & I_{m_2} \end{bmatrix}. \end{aligned}$$

Assume for the moment that A is stable; then from Theorem 2.1 it follows that the Toeplitz operator $\mathcal{R}_c$ is boundedly invertible and $X(\gamma)$ is given by (2.53). From (2.35), (2.36) and (2.37) it follows that in the representation formula (2.53) only $\mathcal{R}_c$ depends on γ. Using (2.37) it follows that the Toeplitz operator $\mathcal{R}_c$ corresponding to the Popov triplet Σ_c has the expression

$$\mathcal{R}_c(\gamma) = \begin{bmatrix} -\gamma^2 I_{m_1} + \mathcal{T}_{11}^* \mathcal{T}_{11} & \mathcal{T}_{11}^* \mathcal{T}_{12} \\ \mathcal{T}_{12}^* \mathcal{T}_{11} & \mathcal{T}_{12}^* \mathcal{T}_{12} \end{bmatrix},$$

where $\mathcal{T}_{11}$ and $\mathcal{T}_{12}$ are the Toeplitz operators associated with T_{11} and T_{12} respectively. Since $\mathcal{T}_{12}^* \mathcal{T}_{12}$ is boundedly invertible as it results from (3.37), $\mathcal{R}_c(\gamma)$ may be rewritten in the equivalent form

$$\begin{aligned} \mathcal{R}_c &= \begin{bmatrix} I & \mathcal{T}_{11}^* \mathcal{T}_{12} (\mathcal{T}_{12}^* \mathcal{T}_{12})^{-1} \\ 0 & I \end{bmatrix} \\ &\times \begin{bmatrix} -\gamma^2 I + \mathcal{T}_{11}^* \mathcal{T}_{11} - \mathcal{T}_{11}^* \mathcal{T}_{12} (\mathcal{T}_{12}^* \mathcal{T}_{12})^{-1} \mathcal{T}_{12}^* \mathcal{T}_{11} & 0 \\ 0 & \mathcal{T}_{12}^* \mathcal{T}_{12} \end{bmatrix} \\ &\times \begin{bmatrix} I & 0 \\ (\mathcal{T}_{12}^* \mathcal{T}_{12})^{-1} \mathcal{T}_{12}^* \mathcal{T}_{11} & I \end{bmatrix}, \end{aligned}$$

which shows that $\mathcal{R}_c^{-1}(\gamma)$ increases with respect to γ and therefore, based on (2.53) we conclude that $X(\gamma)$ is decreasing with respect to γ.

If A is not stable the same development as above can be used for the $\left(0, \tilde{F}\right)$-equivalent Popov triplet $\tilde{\Sigma}_c$, where

$$\tilde{F} = \begin{bmatrix} 0 \\ F_2 \end{bmatrix},$$

with F_2 such that $A + BF_2$ is stable. Then it follows that the stabilizing solution of the ARE$(\tilde{\Sigma}_c)$ is decreasing with respect to γ. Since $\tilde{\Sigma}_c$ is the $\left(0, \tilde{F}\right)$-equivalent with Σ_c, according to (3) from Proposition 2.4 the

ARE($\tilde{\Sigma}_c$) and the ARE(Σ_c) have the same stabilizing solutions and hence, $X(\gamma)$ decreases with respect to γ.

The proof is similar for $Y(\gamma)$. □

A consequence of the lemma above is that $\rho(X(\gamma)Y(\gamma))$ is a decreasing function of γ, and hence the following bisection-type γ-procedure may be used to determine an approximation $\hat{\gamma}_0$ of γ_0 with the assigned level of tolerance $\hat{\epsilon} > 0$:

Step 1 Set $\gamma_L = 0$ and find $\gamma_U > 0$ such that the standard H^∞ problem has a solution for $\gamma = \gamma_U$;

Step 2 Let $\gamma = (\gamma_L + \gamma_U)/2$; if the H^∞-control problem has a solution for γ then set $\gamma_U = \gamma$ and go to Step 3; otherwise, set $\gamma_L = \gamma$ and repeat Step 2;

Step 3 If $\gamma_U - \gamma_L < \hat{\epsilon}$ then set $\hat{\gamma}_0 = \gamma_U$ and STOP; otherwise, return to Step 2.

The γ-procedure above generates two sequences, namely $\{\gamma_{U_k}\}_{k\geq 0}$ and $\{\gamma_{L_k}\}_{k\geq 0}$, converging towards the optimum level of attenuation when k tends to infinity.

If $\lim_{k\to\infty} \gamma_{U_k}^2 - \rho(X(\gamma_{U_k})Y(\gamma_{U_k})) = \delta > 0$ then it follows that the optimum γ_0 corresponds to the case (C1), and in this situation formulae (5.5) give well conditioned computations for any γ in the neighborhood of γ_0. In the situation when $\lim_{k\to\infty} \gamma_{U_k}^2 - \rho(X(\gamma_{U_k})Y(\gamma_{U_k})) = 0$ it results that γ_0 is a solution of the transcendental equation (5.1), in which case the computations determined by formulae (5.5) become sensitive at and near the optimum, since $\gamma_0^2 I - X(\gamma_0)Y(\gamma_0)$ is singular. In the sequel we shall focus our attention on this latter case; therefore the following additional assumption will be introduced:

(A4) γ_0 verifies (5.1) and the Riccati solutions $X(\gamma_0)$ and $Y(\gamma_0)$ are stabilizing and positive definite.

Remark 5.2 Since $\rho(X(\gamma)Y(\gamma))$ is continuous and decreasing with respect to γ, it follows that γ_0 is the unique solution to (5.1).

□

Consider now that the generalized system T having the realization

$$\left(A, \begin{bmatrix} B_1 & B_2 \end{bmatrix}, \begin{bmatrix} C_1 \\ C_2 \end{bmatrix}, \begin{bmatrix} 0 & D_{12} \\ D_{21} & 0 \end{bmatrix}\right), \tag{5.6}$$

is balanced with respect to the solutions $X(\gamma)$ and $Y(\gamma)$ of equations (5.2) and (5.3), respectively, that is

$$X(\gamma) = Y(\gamma) = \begin{bmatrix} s_1(\gamma)I_1 & 0 \\ 0 & S_{22}(\gamma) \end{bmatrix} \tag{5.7}$$

with

$$S_{22}(\gamma) = \text{diag}\,(s_2(\gamma)I_2, ..., s_p(\gamma)I_p)$$

where $s_1(\gamma) > ... > s_p(\gamma) > 0$ and I_k are $n_k \times n_k$ unit matrices, $k = 1, ..., p$. This particular realization of the generalized system does not reduce the generality of the further developments since under the assumption (A3), one may always determine a balanced realization by performing the following procedure, similar to the one described in Section 1.4 for the balancing with respect to the Gramians, namely:

Step 1 Compute the stabilizing solutions $X(\gamma)$ and $Y(\gamma)$ of the AREs (5.2) and (5.3), respectively;

Step 2 Perform the Cholesky factorization $X(\gamma) = R^T(\gamma)R(\gamma)$;

Step 3 Determine the singular value descomposition $R(\gamma)Y(\gamma)R^T(\gamma) = U(\gamma)\Sigma^2(\gamma)U^T(\gamma)$ with $U(\gamma)$ orthogonal;

Step 4 Define the balancing transformation $T_b(\gamma) = \Sigma^{-\frac{1}{2}}(\gamma)U^T(\gamma)R(\gamma)$ and compute the equivalent realization of the generalized system to obtain the balanced realization with respect to the Riccati solutions

$$\left(T_bAT_b^{-1}, T_b\begin{bmatrix} B_1 & B_2 \end{bmatrix}, \begin{bmatrix} C_1 \\ C_2 \end{bmatrix}T_b^{-1}, \begin{bmatrix} 0 & D_{12} \\ D_{21} & 0 \end{bmatrix}\right).$$

In order to simplify the notation we shall adopt the notation (5.6) for this balanced realization.

From the procedure above it follows that all matrices in the balanced realization are functions of γ; these dependences are smooth for $\gamma \geq \gamma_0$ since the Riccati solutions $X(\gamma)$ and $Y(\gamma)$ are smooth, which fact may be easily deduced by an implicit function argument together with the uniqueness of the stabilizing solutions to the Riccati equations. Indeed, let $X(\tilde{\gamma})$ be the stabilizing solution of (5.2) for a certain $\tilde{\gamma} \geq \gamma_0$ and denote

$$\mathcal{F}(X, \gamma) := A^TX + XA + X\left(\gamma^{-2}B_1B_1^T - B_2B_2^T\right)X + C_1^TC_1.$$

Therefore, we have $\mathcal{F}(\widetilde{X}, \tilde{\gamma}) = 0$ and

$$\begin{aligned}(\partial_X\mathcal{F})(\widetilde{X}, \tilde{\gamma})\,(X) &= \left[A + \tilde{\gamma}^{-2}(B_1B_1^T - B_2B_2^T)\widetilde{X}\right]^T X \\ &\quad + X\left[A + \tilde{\gamma}^{-2}(B_1B_1^T - B_2B_2^T)\widetilde{X}\right],\end{aligned}$$

which is invertible since $A + \tilde{\gamma}^{-2}(B_1B_1^T - B_2B_2^T)\widetilde{X}$ is stable and X is positive definite. Then according to the Implicit Function Theorem there exist a neighborhood of $\tilde{\gamma}$ and a smooth function $X(\gamma)$ defined in this neighborhood, such that $\mathcal{F}(X, \gamma) = 0$ and $X(\gamma)$ is close to $X(\tilde{\gamma})$ for γ close to γ_0;

hence, it follows that $A + \left(\gamma^{-2}B_1B_1^T - B_2B_2^T\right)X(\gamma)$ is stable for γ close to $\tilde{\gamma}$, and from the uniqueness of the stabilizing solution we conclude that $X(\gamma)$ is the stabilizing solution of the Riccati equation which is smooth with respect to γ. A similar argumentation is available for the smoothness of $Y(\gamma)$.

Further, taking into account the assumptions (A3) and (A4) it results that all matrices from the balanced realization (5.6) with respect to the Riccati solutions are smooth for $\gamma \geq \gamma_0$.

Since we are interested in the behavior of the solution to the H^∞ problem near the optimum γ_0, in the sequel we shall investigate the case where γ tends to γ_0. Based on the smoothness of the Riccati solutions with respect to γ it follows that the function

$$\epsilon(\gamma) = \gamma^2 - \rho\left(X(\gamma)(Y(\gamma))\right)$$

is smooth and $\epsilon(\gamma) \to 0$ when $\gamma \to \gamma_0$, and taking into account (5.4) and (5.7) it results that

$$\epsilon(\gamma) = \gamma^2 - s_1^2(\gamma).$$

In the following developments we shall omit writing the explicit dependence upon γ of the matrices involved, but we shall keep in mind this dependence. For the balanced realization considered, the matrix Z becomes

$$Z = \begin{bmatrix} \frac{s_1(s_1^2+\epsilon)}{\epsilon} I_1 & 0 \\ 0 & Z_{22} \end{bmatrix}, \tag{5.8}$$

where

$$Z_{22} := (s_1^2+\epsilon)S_{22}\left[(s_1^2+\epsilon)^2 I - S_{22}^2\right]^{-1}.$$

Further, perform the following partitions conformably with (5.8),

$$\begin{aligned} A &= \begin{bmatrix} A_{11} & A_{12} \\ A_{21} & A_{22} \end{bmatrix}, \\ B_1 &= \begin{bmatrix} B_{11} \\ B_{21} \end{bmatrix}, \qquad B_2 = \begin{bmatrix} B_{12} \\ B_{22} \end{bmatrix}, \\ C_1 &= [C_{11}\; C_{12}], \qquad C_2 = [C_{21}\; C_{22}]. \end{aligned} \tag{5.9}$$

Then, using (5.8) and (5.9), formulae (5.5) give

$$\begin{aligned} A_K &= \begin{bmatrix} A_{K_{11}} & A_{K_{12}} \\ A_{K_{21}} & A_{K_{22}} \end{bmatrix}, \\ B_K &= \begin{bmatrix} s_1^3 C_{21}^T \\ Z_{22}C_{22}^T \end{bmatrix}, \\ C_K &= -\begin{bmatrix} s_1 B_{12}^T & B_{22}^T S_{22} \end{bmatrix}, \end{aligned} \tag{5.10}$$

with

$$\begin{aligned}
A_{K_{11}} &:= A_{11} + s_1\left[(s_1^2+\epsilon)^{-1}B_{11}B_{11}^T - B_{12}B_{12}^T\right] - \frac{s_1(s_1^2+\epsilon)C_{21}^TC_{21}}{\epsilon},\\
A_{K_{12}} &:= A_{12} + \left[(s_1^2+\epsilon)^{-1}B_{11}B_{21}^T - B_{12}B_{22}^T\right]S_{22} - \frac{s_1(s_1^2+\epsilon)C_{21}^TC_{22}}{\epsilon},\\
A_{K_{21}} &:= A_{21} + s_1\left[(s_1^2+\epsilon)^{-1}B_{21}B_{11}^T - B_{22}B_{12}^T\right] - Z_{22}C_{22}^TC_{21},\\
A_{K_{22}} &:= A_{22} + \left[(s_1^2+\epsilon)^{-1}B_{21}B_{21}^T - B_{22}B_{22}^T\right]S_{22} - Z_{22}C_{22}^TC_{22}.
\end{aligned}$$

The above formulae show that this H^∞ suboptimal controller is in fact a two-time-scale system which fast component is given by the first row of A_K and B_K. According to the theory of singular perturbations this fast dynamics may be reduced if $C_{21}^TC_{21}$ is nonsingular, obtaining thus the reduced-order controller

$$K_0 = \left[\begin{array}{c|c} A_0 & B_0 \\ \hline C_0 & D_0 \end{array}\right],$$

where

$$\begin{aligned}
A_0 &= A_{22} + \left(\gamma_0^{-2}B_{21}B_{21}^T - B_{22}B_{22}^T\right)S_{22} - B_0C_{22},\\
B_0 &= \left(A_{21} + \gamma_0^{-1}B_{21}B_{11}^T - \gamma_0B_{22}B_{12}^T - Z_{22}C_{22}^TC_{21}\right)\\
&\quad \times \left(C_{21}^TC_{21}\right)^{-1}C_{21}^T + Z_{22}C_{22}^T, \qquad (5.11)\\
C_0 &= -D_0C_{22} - B_{22}^TS_{22},\\
D_0 &= -\gamma_0B_{12}^T\left(C_{21}^TC_{21}\right)^{-1}C_{21}^T.
\end{aligned}$$

Remark 5.3 If $C_{21} = 0$ an optimal solution of the H^∞-control problem can be determined by setting $\epsilon = 0$, since in this case no singularity appears in the expressions for A_K and B_K. If $C_{21}^TC_{21}$ is singular one may perform an additional orthogonal transformation to the controller (5.10) such that

$$U^TC_{21}^T(\gamma_0)C_{21}(\gamma_0)U = \begin{bmatrix} \overline{C}_{21}^T(\gamma_0)\overline{C}_{21}(\gamma_0) & 0 \\ 0 & 0 \end{bmatrix},$$

where $\overline{C}_{21}^T\overline{C}_{21}$ is nonsingular. Thus we obtain again a singularly perturbed system whose fast component has dimension equal to the rank of $C_{21}^TC_{21}$ and which may be reduced according to the theory of singular perturbations; we conclude that if n denotes the dimension of A then the order of the reduced controller equals $n - \text{rank}(C_{21}^TC_{21})$. □

The main result of this section, providing an optimal solution to the standard H^∞ problem, is given by the following theorem.

Theorem 5.1 *If $C_{21}^T(\gamma_0)C_{21}(\gamma_0)$ is nonsingular then the controller (5.11) is generically an optimal solution of the standard H^∞-control problem.*

Remark 5.4 The term 'generic' in the statement will be explained at the end of the proof, where a stabilizability property is assumed. □

Proof. We have to prove that the controller (5.11) is stabilizing and the H^∞-norm of the resulting system is equal to γ_0. According to Proposition 3.3 a controller is a solution to the H^∞ problem for the generalized system T if and only if it is also a solution to the H^∞ problem corresponding to the system T_o defined by (3.67), which becomes, under the assumption of the normalizing conditions (3.52),

$$\begin{aligned} \dot{x} &= (A+B_1F_1)\,x + B_1u_1 + B_2u_2, \\ y_1 &= -F_2x + u_2, \\ y_2 &= C_2x + D_{21}u_1, \end{aligned} \tag{5.12}$$

where

$$\begin{aligned} F_1 &:= \gamma^{-2}B_1X, \\ F_2 &:= -B_2^TX. \end{aligned}$$

It results that the suboptimal controller (5.5) stabilizes (5.12) and the resulting system has the H^∞-norm less than γ. The controller (5.5) may be written in the equivalent form

$$\begin{aligned} \dot{x}_K &= (A+B_1F_1+B_2F_2+H_2C_2)\,x_K - H_2y_2, \\ u_2 &= F_2x_K, \end{aligned} \tag{5.13}$$

where

$$H_2 := -ZC_2^T.$$

When coupling the controller (5.13) to the system (5.12), one obtains the realization (A_R, B_R, C_R, D_R) of the resulting system with

$$\begin{aligned} A_R &= \begin{bmatrix} A+B_1F_1 & B_2F_2 \\ -H_2C_2 & A+B_1F_1+B_2F_2+H_2C_2 \end{bmatrix}, \\ B_R &= \begin{bmatrix} B_1 \\ -H_2D_{21} \end{bmatrix}, \\ C_R &= \begin{bmatrix} -F_2 & F_2 \end{bmatrix}, \\ D_R &= 0. \end{aligned}$$

An equivalent realization of the resulting system above is given by performing the nonsingular transformation

$$S = \begin{bmatrix} I & 0 \\ -I & I \end{bmatrix},$$

thus obtaining

$$\begin{aligned}
\widetilde{A}_R &= SA_RS^{-1} = \begin{bmatrix} A + B_1F_1 + B_2F_2 & B_2F_2 \\ 0 & A + B_1F_1 + H_2C_2 \end{bmatrix}, \\
\widetilde{B}_R &= SB_R = \begin{bmatrix} B_1 \\ -(B_1 + H_2D_{21}) \end{bmatrix}, \\
\widetilde{C}_R &= C_RS^{-1} = \begin{bmatrix} 0 & F_2 \end{bmatrix}
\end{aligned}$$

where $A+B_1F_1+B_2F_2$ is stable, since X is a stabilizing solution of the ARE (5.2). This equivalent realization shows that the resulting system includes an unobservable part which dynamics is determined by $A + B_1F_1 + B_2F_2$; after the removal of this unobservable stable part we deduce that the resulting system has the following state space representation

$$\begin{aligned}
\dot{\tilde{x}} &= (A + B_1F_1 + H_2C_2)\,\tilde{x} - (B_1 + H_2D_{21})\,u_1, \qquad (5.14)\\
y_1 &= F_2\tilde{x}.
\end{aligned}$$

This system is stable and its H^∞-norm is less than γ. If γ is close to γ_0 and the realization of the generalized system T is balanced with respect to the Riccati solutions, formulae (5.14) together with (5.7), (5.8) and (5.9), give

$$\begin{aligned}
\dot{\tilde{x}}_1 &= \left(A_{11} + s_1^{-1}B_{11}B_{11}^T - \frac{s_1^3}{\epsilon}C_{21}^TC_{21}\right)\tilde{x}_1 \\
&\quad + \left(A_{12} + s_1^{-2}B_{11}B_{12}^TS_{22} - \frac{s_1^3}{\epsilon}C_{21}^TC_{22}\right)\tilde{x}_2 - \left(B_{11} - \frac{s_1^3}{\epsilon}C_{21}^TD_{21}\right)u_1, \\
\dot{\tilde{x}}_2 &= \left(A_{21} + s_1^{-1}B_{21}B_{11}^T - Z_{22}S_{22}C_{22}^TC_{21}\right)\tilde{x}_1 \\
&\quad + \left(A_{22} + s_1^{-2}B_{21}B_{21}^TS_{22} - Z_{22}S_{22}C_{22}^TC_{22}\right)\tilde{x}_2 \\
&\quad - \left(B_{21} - Z_{22}S_{22}C_{22}^TD_{21}\right)u_1, \qquad (5.15)\\
y_1 &= -s_1B_{12}^T\tilde{x}_1 - B_{22}^TS_{22}\tilde{x}_2,
\end{aligned}$$

where we have partitioned $\tilde{x}$ in conformity with (5.8),

$$\tilde{x} := \begin{bmatrix} \tilde{x}_1 \\ \tilde{x}_2 \end{bmatrix}.$$

Equations (5.15) show that near the optimum the resulting system has two-time-scale dynamics, which fast component corresponding to the state $\tilde{x}_1$ may be reduced if $C_{21}^T C_{21}$ is invertible. Then the slow component of the resulting system (5.15) is given by

$$\begin{aligned} \dot{\tilde{x}}_2 &= \widetilde{A}\tilde{x}_2 + \widetilde{B}u_1 \\ y_1 &= \widetilde{C}\tilde{x}_2 + \widetilde{D}u_1 \end{aligned} \tag{5.16}$$

where

$$\begin{aligned} \widetilde{A} &= A_{22} + \gamma_0^{-2} B_{21} B_{21}^T S_{22} - Z_{22} C_{22}^T C_{22} \\ &\quad - \left(A_{21} + \gamma_0^{-1} B_{21} B_{11}^T - Z_{22} C_{22}^T C_{21}\right) \left(C_{21}^T C_{21}\right)^{-1} C_{21}^T C_{22}, \\ \widetilde{B} &= -B_{21} + Z_{22} C_{22}^T D_{21} + \left(A_{21} + \gamma_0^{-1} B_{21} B_{11}^T - Z_{22} C_{22}^T C_{21}\right) \\ &\quad \times \left(C_{21}^T C_{21}\right)^{-1} C_{21}^T D_{21}, \\ \widetilde{C} &= \gamma_0 B_{12}^T \left(C_{21}^T C_{21}\right)^{-1} C_{21}^T C_{22} - B_{22}^T S_{22}, \\ \widetilde{D} &= -\gamma_0 B_{12}^T \left(C_{21}^T C_{21}\right)^{-1} C_{21}^T D_{21}. \end{aligned} \tag{5.17}$$

Direct algebraic computations show that the system above coincides with the resulting system obtained by coupling the optimal controller (5.11) to the system (5.12). Therefore we have to prove that the system (5.16) is stable and its H^∞-norm equals γ_0. To this end consider the norm type of Riccati inequality associated with the system $\left(\widetilde{A}^T, \widetilde{C}^T, \widetilde{B}^T, \widetilde{D}^T\right)$, namely,

$$\begin{aligned} \widetilde{A}\Pi + \Pi\widetilde{A}^T + \left(\Pi\widetilde{C}^T + \widetilde{B}\widetilde{D}^T\right)\left(\tilde{\gamma}^2 I - \widetilde{D}\widetilde{D}^T\right)^{-1} \\ \times \left(\widetilde{C}\Pi + \widetilde{D}\widetilde{B}^T\right) + \widetilde{B}\widetilde{B}^T \leq 0. \end{aligned} \tag{5.18}$$

We show now that for any $\tilde{\gamma} > \gamma_0$ this inequality is verified by $Z_{22}(\gamma_0) > 0$. Indeed, in Section 3.4 it has been shown that a consequence of Theorem 3.5 for the case where the normalizing conditions (3.52) hold, is that the ARE($\Sigma_\times$) (3.78) has a stabilizing solution; in the unscaled form considered ($\gamma \neq 1$) this equation becomes

$$\begin{aligned} &(A + B_1 F_1) Z + Z (A + B_1 F_1)^T \\ &+ Z\left(\gamma^{-2} F_2^T F_2 - C_2^T C_2\right) Z + B_1^T B_1 = 0, \end{aligned} \tag{5.19}$$

and according to Theorem 3.9 its stabilizing solution is

$$Z = \gamma^2 Y \left(\gamma^2 I - XY\right)^{-1}.$$

When writing (5.19) in the partitioned form corresponding to (5.8), one obtains:

- The block (1,1) of (5.19):

$$\frac{s_1^3}{\epsilon}\left(A_{11}+s_1^{-1}B_{11}B_{11}^T\right)+\frac{s_1^3}{\epsilon}\left(A_{11}+s_1^{-1}B_{11}B_{11}^T\right)^T+\frac{s_1^6}{\epsilon^2}B_{12}B_{12}^T$$
$$-\frac{s_1^6}{\epsilon^2}C_{21}^TC_{21}+B_{11}B_{11}^T=0$$

For $\epsilon \to 0$ it results that

$$B_{12}B_{12}^T=C_{21}^TC_{21}. \tag{5.20}$$

- The block (1,2) of (5.19)

$$s_1^2\left(A_{12}+s_1^{-2}B_{11}B_{21}^TS_{22}\right)\left(s_1^2I-S_{22}^2\right)^{-1}S_{22}+\frac{s_1^3}{\epsilon}\left(A_{21}^T+s_1^{-1}B_{11}B_{21}^T\right)$$
$$+\frac{s_1^4}{\epsilon}B_{12}B_{22}^TS_{22}^2(s_1^2I-S_{22}^2)^{-1}-\frac{s_1^5}{\epsilon}C_{21}^TC_{22}S_{22}\left(s_1^2I-S_{22}^2\right)^{-1}$$
$$+B_{11}B_{21}^T=0,$$

which gives, for $\epsilon \to 0$,

$$\left(A_{21}+s_1^{-1}B_{21}B_{11}^T\right)^T-C_{21}^TC_{22}Z_{22}=-s_1^{-1}B_{12}B_{22}^TS_{22}Z_{22}. \tag{5.21}$$

- The block (2,2) of (5.19)

$$\left(A_{22}+s_1^{-2}B_{21}B_{21}^TS_{22}\right)Z_{22}+Z_{22}\left(A_{22}+s_1^{-2}B_{21}B_{21}^TS_{22}\right)^T$$
$$+\gamma^{-2}Z_{22}S_{22}B_{22}B_{22}^TS_{22}Z_{22}-Z_{22}C_{22}^TC_{22}Z_{22}+B_{21}B_{21}^T$$
$$=0. \tag{5.22}$$

Using (5.21) we obtain the equivalent expression of $\widetilde{A}$ defined in (5.17)

$$\begin{aligned}\widetilde{A} &= A_{22}+s_1^{-2}B_{21}B_{21}^TS_{22}-Z_{22}C_{22}^TC_{22}\\ &\quad +s_1^{-1}Z_{22}S_{22}B_{22}B_{12}^T\left(C_{21}^TC_{21}\right)^{-1}C_{21}^TC_{22}.\end{aligned} \tag{5.23}$$

Then, based on (5.20) and (5.22), one may directly verify that the Riccati inequality (5.18) is verified by $\Pi = Z_{22}(\gamma_0) > 0$, which fact implies that

$$\widetilde{A}Z_{22}(\gamma_0)+Z_{22}(\gamma_0)\widetilde{A}^T+\widetilde{B}\widetilde{B}^T \le 0.$$

Assuming generically that the pair $\left(\widetilde{A},\widetilde{B}\right)$ is stabilizable, from the above inequality it follows that $\widetilde{A}$ is stable, and hence the controller (5.11) is stabilizing.

On the other hand, since in the Riccati inequation (5.18) $\widetilde{A}$ is stable, according to the Bounded Real Lemma (Corollary 2.7) it results that the system with the realization $\left(\widetilde{A}^T, \widetilde{C}^T, \tilde{B}^T, \widetilde{D}^T\right)$ has H^∞-norm less than or equal to $\tilde{\gamma}$, and hence the same is true for $\left(\widetilde{A}, \widetilde{B}, \tilde{C}, \widetilde{D}\right)$. Then, taking into account that $\tilde{\gamma} > \gamma_0$ is arbitrarily chosen, it follows that the H^∞-norm of the resulting system (5.17) is just γ_0, and the proof is concluded. □

Summarizing the results above, we propose the following well conditioned procedure to determine under the assumptions (A1)–(A4), an optimal solution to the H^∞ problem.

Step 1 Compute the optimal level of attenuation γ_0 by using the bisection based γ-procedure described at the beginning of this section;

Step 2 Determine a balanced realization of the generalized system T with respect to the Riccati solutions $X(\gamma_0)$ and $Y(\gamma_0)$;

Step 3 Using the elements of the balanced realization, determine the optimal controller given by formulae (5.11);

Step 4 (Testing the controller) Couple the controller determined at Step 3, to the generalized system T and compute the eigenvalues of the resulting system and its H^∞ -norm.

Further, we consider the case when the orthogonality conditions in (A2) do not hold and (A3) is replaced by the regularity assumptions (R1) and (R2) adopted in Section 3.1. In this case the two Riccati equations associated to the H^∞-control problem are

$$\begin{aligned} &A^T X + XA + \gamma^{-2} X B_1 B_1^T X \\ &\quad - \left(XB_2 + C_1^T D_{12}\right)\left(B_2^T X + D_{12}^T C_1\right) + C_1^T C_1 = 0, \qquad (5.24) \\ &AY + YA^T + \gamma^{-2} Y C_1^T C_1 Y \\ &\quad - \left(YC_2^T + B_1 D_{21}^T\right)\left(C_2 Y + D_{21} B_1^T\right) + B_1 B_1^T = 0, \qquad (5.25) \end{aligned}$$

and the suboptimal H^∞-controller is given by

$$\begin{aligned} \dot{x}_K &= \left[A + B_1 F_1 + B_2 F_2 + H_2 \left(C_2 + D_{21} F_1\right)\right] x_K - H_2 y_2, \\ u_2 &= F_2 x_K, \end{aligned}$$

where

$$H_2 := -\left[Z\left(C_2 + D_{21} F_1\right)^T + B_1 D_{21}^T\right].$$

Since the realization (A, B_1, C_1) is no more minimal, one may obtain the Riccati solutions $X(\gamma_0) \geq 0$ and (or) $Y(\gamma_0) \geq 0$, and hence the balancing procedure described above is no longer valid in this situation. We shall then use the balancing procedure with respect to positive semi-definite matrices described in (Glover, 1984) which states that there exists a nonsingular transformation M such that

$$\begin{aligned} MXM^T &= \begin{bmatrix} X_{11} & 0 \\ 0 & X_{22} \end{bmatrix}, \\ \left(M^{-1}\right)^T YM^{-1} &= \begin{bmatrix} Y_{11} & 0 \\ 0 & Y_{22} \end{bmatrix}, \end{aligned}$$

with $X_{11} = Y_{11} = \operatorname{diag}(\sigma_1 I_1, ..., \sigma_r I_r)$ where $\sigma_1 > ... > \sigma_r > 0$, I_i are $n_i \times n_i$ unit matrices, $i = 1, ..., r$, $X_{22} \geq 0$, $Y_{22} \geq 0$ and $X_{22}Y_{22} = 0$.

By performing this balanced transformation to the generalized system, the solutions of the Riccati equations (5.24) and (5.25) become

$$X = \begin{bmatrix} s_1 I_1 & 0 \\ 0 & \widetilde{X}_{22} \end{bmatrix}, \quad Y = \begin{bmatrix} s_1 I_1 & 0 \\ 0 & \widetilde{Y}_{22} \end{bmatrix},$$

where $\widetilde{X}_{22} := \operatorname{diag}(\sigma_2 I_2, ..., \sigma_r I_r, X_{22})$ and $\widetilde{Y}_{22} := \operatorname{diag}(\sigma_2 I_2, ..., \sigma_r I_r, Y_{22})$.

Since $\sigma_1 > ... > \sigma_r$ and $X_{22}Y_{22} = 0$, it follows that the condition $\gamma_0^2 - \rho(X(\gamma_0)Y(\gamma_0)) = 0$ is equivalent to $s_1^2 = \rho(XY)$, therefore we may determine an optimal H^∞ controller using the same steps as in the case of the standard H^∞-control problem considered above. Hence, after determining a balanced realization with respect to $X(\gamma_0)$ and $Y(\gamma_0)$ in the sense above, we obtain the following realization of the optimal H^∞ controller, generalizing (5.11)

$$\begin{aligned} A_0 &= A_{22} + B_{21}F_{12} + B_{22}F_{22} - B_0\widetilde{C}_{22}, \\ B_0 &= \left[A_{21} + B_{21}F_{11} + B_{22}F_{21} - \left(Z_{22}\widetilde{C}_{22}^T + B_{21}D_{21}^T\right)\widetilde{C}_{21}\right] \\ &\quad \times \left(\widetilde{C}_{21}^T\widetilde{C}_{21}\right)^{-1}\widetilde{C}_{21}^T + Z_{22}\widetilde{C}_{22}^T + B_{21}D_{21}^T, \qquad (5.26) \\ C_0 &= -D_0\widetilde{C}_{22} + F_{22}, \\ D_0 &= F_{21}\left(\widetilde{C}_{21}^T\widetilde{C}_{21}\right)^{-1}\widetilde{C}_{21}^T, \end{aligned}$$

where

$$\begin{aligned} F_1 &:= \gamma_0^{-2}B_1^T X = \left[\, F_{11} \quad F_{12} \,\right], \\ F_2 &:= -\left(B_2^T X + D_{12}^T C_1\right) = \left[\, F_{21} \quad F_{22} \,\right], \\ \widetilde{C}_{21} &:= C_{21} + D_{21}F_{11}, \\ \widetilde{C}_{22} &:= C_{22} + D_{21}F_{12}, \\ Z_{22} &:= \gamma_0^2\widetilde{Y}_{22}\left(\gamma_0^2 I - \widetilde{X}_{22}\widetilde{Y}_{22}\right)^{-1}, \end{aligned}$$

in which $\widetilde{C}_{21}^T\widetilde{C}_{21}$ has been assumed nonsingular.

5.2. CASE STUDIES

Case study 1: A model matching problem

Consider the same design problem as in Case study 2 of Section 4.1.3, in which only the nominal flight conditions of the F-4E fighter are changed, corresponding to subsonic flight, namely Mach 0.85 and altitude 5000 ft. The aerodynamic data from the state equations (4.67) are in this case $Z_\alpha = -1.702\,\text{rad/s}$, $Z_\delta = 0.481\,\text{rad/s}$, $M_\alpha = 11.163\,\text{rad/s}^2$, $M_q = -1.418\,\text{rad/s}$, $M_\delta = -36.269\,\text{rad/s}^2$, $\tau = 1/14\,\text{s}$, $n_\alpha = 50.72\,\text{g/rad}$, $n_\delta = -19.433\,\text{g/rad}$ (Cavallo *et al.*, 1992). The open loop eigenvalues are $\{-4.9091; 1.7841; -14\}$ showing that the short period dynamics is unstable at the nominal flight conditions under consideration; it is easy to check that (DF2) from the specific assumptions to the DFP is no more valid in this case and hence, the equivalent H^∞ problem formulated in Section 4.1.3 for the generalized system (4.69), (4.70) must be solved without reducing it to a DFP.

The same numerical values have been used for the weighting functions W, V and V_n as the ones considered in case study Section 4.1.3. Based on the γ-procedure described in Section 5.1 we computed the optimal level of attenuation $\gamma_0 = 0.6668$ which verifies the transcendental equation (5.1). The normalizing conditions (3.52) are not fulfilled since $D_{21}^T B_1 \neq 0$ and then we used formulae (5.26) to determine the optimal solution to the H^∞-control problem. Thus we obtained the reduced order optimal controller

$$K_0 = \left[\begin{array}{c|c} A_0 & B_0 \\ \hline C_0 & D_0 \end{array}\right],$$

where

$$\begin{aligned}
A_0 &= \begin{bmatrix} 3.5752 & 1.4050 & -0.2451 & 0.3043 & -0.3901 \\ -29.1119 & -7.2999 & -1.1971 & -0.5585 & 0.3940 \\ -40.5264 & -3.9509 & 0.9612 & -3.1774 & 3.3839 \\ 60.0125 & 17.5200 & -5.3513 & -6.8698 & -19.6757 \\ 41.0656 & 10.8198 & -6.8945 & 9.1881 & -10.9113 \end{bmatrix}, \\
B_0 &= \begin{bmatrix} 0.2563 & -0.0179 & -0.0012 \\ 0.7262 & 0.0201 & 0.0014 \\ -2.2164 & 0.1594 & 0.0110 \\ 7.6014 & -0.5665 & -0.0389 \\ 4.9850 & -0.3676 & -0.0253 \end{bmatrix}, \\
C_0 &= \begin{bmatrix} -5.8069 & -1.4836 & 0.8159 & -0.8168 & 1.0867 \end{bmatrix}, \\
D_0 &= \begin{bmatrix} -0.6668 & 0.0496 & 0.0034 \end{bmatrix}.
\end{aligned}$$

Then we solved the same H^∞ problem by using the suboptimal controllers given by (3.105) (adapted to the case $\gamma \neq 1$) for $\gamma_1 = 0.667$ and $\gamma_2 = 1$. The time responses of the tracking error $e(t)$ and of the elevator deflection $\delta_e(t)$ the step reference input $\alpha_{ref} = 0.2\,\text{rad}$ are plotted in Figures 5.1 and 5.2 respectively, showing that the results are almost identical for γ_0 and γ_1; it can be also seen that the improvement of the tracking performances corresponding to γ_0 and γ_1 does not require a significant increasing of the control effort δ_e.

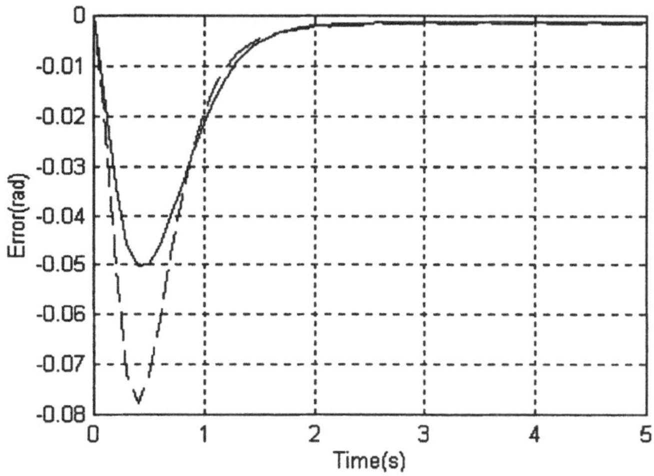

Figure 5.1. The tracking error response

We also considered the following criteria in order to compare the optimal and the suboptimal H^∞ controllers:

1) The value of $\|T_{y_1u_1}\|_\infty$ which reveals the attenuation properties ensured by the controller.

2) The largest absolute value of the elements of matrices giving the realization of the controllers; we denoted this value by $M_{\max}$.

3) The *relative internal stability margin* S_{mar} of the state matrix of the resulting system

$$A_R = \begin{bmatrix} A + B_2D_KC_2 & B_2C_K \\ B_KC_2 & A_K \end{bmatrix},$$

defined as

$$S_{\text{mar}} := \frac{r(A_R)}{\|A_R\|},$$

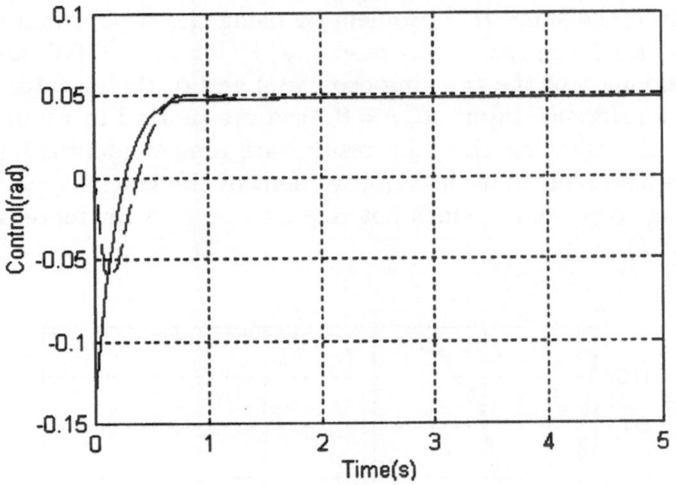

Figure 5.2. The control response

TABLE 5.1 Comparative results

γ	0.6668 *optimal*	0.6670	1
K order	5	6	6
$\Vert T_{y_1u_1}\Vert_\infty$	0.6668	0.6669	0.8634
$M_{\max}$	60.01	$4.32.10^5$	648.15
S_{mar}	4.10^{-4}	6.10^{-8}	3.10^{-5}

where the complex stability radius of the stable matrix A_R is given by

$$r(A_R) = \left(\left\|(sI - A_R)^{-1}\right\|_\infty\right)^{-1}.$$

The obtained results are given in Table 5.1, pointing out a significant degradation of $M_{\max}$ and S_{mar} when a suboptimal H^∞ controller near the optimum is used, confirming the ill-conditioned computation determined by the formulae corresponding to the suboptimal case for γ close to γ_0.

The robustness properties given by the *gain margin* (GM) and the *phase margin* (PM) have been also investigated, by considering the multiplicative unstructured uncertainty in the dynamics of the actuator and of sensors, respectively. Then GM= $[1 - r_{\min}, 1 + r_{\min}]$, PM= $[-\theta, \theta]$ where $r_{\min} = (\| GK(I + GK)^{-1} \|_\infty)^{-1}$ for output uncertainty (at sensors) and $r_{\min} = (\| K(I + GK)^{-1}G \|_\infty)^{-1}$ for input uncertainty (at the actuator). These stability margins are given in Table 5.2.

TABLE 5.2 Stability margins GM(dB), PM(deg)

γ	0.6668 *optimal*	0.6670	1
GM_o	[−5.0161; 3.1594]	[−4.8932; 3.1110]	[−4.0574; 2.7547]
GM_i	[−4.9023; 3.1146]	[−4.9008; 3.1140]	[−4.0657; 2.7585]
PM_o	25.3417	24.8721	21.5089
PM_i	24.9073	24.9015	21.5438

Although these results are not very satisfactory (in fact the robustness performances have not been explicitly formulated in the design specifications), the best values are achieved at and near the optimum.

Case study 2: Optimal robust design with respect to additive uncertainty

Given a nominal system G and $\delta > 0$, the robust design problem with respect to additive uncertainty consists in determining a controller which stabilizes all systems $G_\Delta = G + \Delta$ with $\Delta \in \mathrm{RH}_+^\infty$ and $\|\Delta\|_\infty \leq \delta$. The problem is illustrated in Figure 5.3 from which it follows that the input-output dependence from v to w is given by

$$T_{wv} = K(I - GK)^{-1}.$$

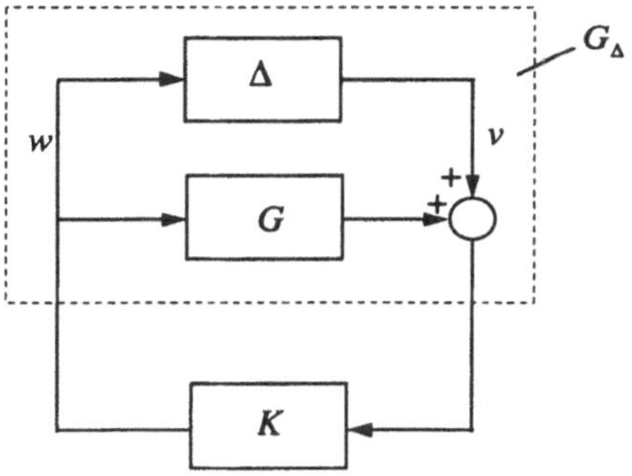

Figure 5.3. The additive uncertainty modeling

Then, based on the Small Gain Theorem (Corollary 2.8), one obtains:

Proposition 5.1 *If K stabilizes G and*

$$\left\|K(I - GK)^{-1}\right\|_\infty < \gamma := \delta^{-1} \tag{5.27}$$

then K is stabilizing for all systems $G_\Delta = G + \Delta$ with $\Delta \in \mathrm{RH}_+^\infty$ and $\|\Delta\|_\infty \leq \delta$. □

The sufficient conditions given by the proposition above can be accomplished by solving the H^∞- control problem formulated for the generalized system

$$T = \begin{bmatrix} 0 & I \\ I & G \end{bmatrix}. \tag{5.28}$$

Indeed, one can directly check that LFT(T, K) (see (3.11) and (3.12)) coincides with the left hand side of (5.27). Notice that if G is stable then the problem is irrelevant since G_Δ is stable for all $\Delta \in \mathrm{RH}_+^\infty$ and hence no controller is required for stabilizing G_Δ.

Although the additive modeling of the dynamic uncertainty is less used than the multiplicative and the coprime factorization based representation of the uncertainty, there are situations when it is important to determine the maximum allowable additive uncertainty for which the design of a robust stabilizing controller is possible.

From (5.28) it follows that $B_1 = 0$ and $C_1 = 0$ in the generalized system T, which shows that if A has eigenvalues on the $j\omega$-axis then the regularity assumptions (R1) and (R2) considered in Section 3.1 are not fulfilled, and hence the robust design based on the H^∞ approach is no more valid.

We shall illustrate the above considerations by a numerical example in which the nominal system G is the third-order short period dynamics of the F-4E fighter considered in the previous case study, with the realization (A, B, C), where

$$\begin{aligned} A &= \begin{bmatrix} -1.7020 & 1 & 0.4810 \\ 11.1630 & -1.4180 & 36.2690 \\ 0 & 0 & -14 \end{bmatrix}, \\ B &= \begin{bmatrix} 0 \\ 0 \\ 14 \end{bmatrix}, \\ C &= \begin{bmatrix} 50.7200 & 0 & -19.4330 \\ 0 & 1 & 0 \end{bmatrix}. \end{aligned} \tag{5.29}$$

The γ-procedure described in Section 5.1 gives $\gamma_0 = 0.0152$ which satisfies the transcendental equation (5.1). Further, we determined by formulae (5.26) the second-order optimal controller $K_0 = \left[\begin{array}{c|c} A_0 & B_0 \\ \hline C_0 & D_0 \end{array}\right]$, where

$$A_0 = \begin{bmatrix} -10.7527 & -12.7326 \\ 3.5589 & -7.5834 \end{bmatrix},$$

$$\begin{aligned} B_0 &= \begin{bmatrix} 0.1376 & 0.0095 \\ 0.1624 & 0.0112 \end{bmatrix}, \\ C_0 &= \begin{bmatrix} 0.3439 & -0.2381 \end{bmatrix}, \\ D_0 &= \begin{bmatrix} 0.0152 & 0.0010 \end{bmatrix}. \end{aligned}$$

In order to check these results we coupled the controller K_0 to the generalized system (5.28), obtaining thus a stable resulting system which H^∞-norm equals γ_0. Let us finally remark that no ill-conditioned computations appeared when determining the above optimal solution to the robust design problem.

NOTES AND REFERENCES

Optimal state space solutions of the H^∞-control problem can be determined using the descriptor representation of the controller together with an all-pass embedding technique as shown in (Safonov *et al.*, 1990) and (Glover *et al.*, 1991). A different method has been proposed in (Gahinet, 1992a) in which the strictly proper suboptimal H^∞ central controller is replaced by proper controllers, and by an appropriate choice of the feedthrough gain D_K the unbounded modes appearing near the optimum are cancelled. The method described in this chapter was first presented in (Dragan *et al.*, 1997) and numerical examples are also given in (Stoica, 1997) and (Dragan and Stoica 1997).

Alternative proofs for the result stated by Lemma 5.1 may be found in (Wimmer, 1985) and (Gahinet, 1992b).

CHAPTER 6

SINGULAR H^∞ PROBLEMS

This chapter deals with two H^∞ optimization problems frequently arising in control applications, namely the robust design with respect to multiplicative uncertainty and the sensitivity reduction. Both problems have been intensively investigated (details are given in Notes and References), but in the present chapter we shall treat them by a technique based on LMIs. Explicit formulae for the corresponding solutions are derived in terms of LMIs solutions; we also discuss numerical aspects related to the solutions computation illustrated by design examples from aircraft domain.

6.1. ROBUSTNESS WITH RESPECT TO MULTIPLICATIVE UNCERTAINTY

The robust design problem with respect to multiplicative uncertainty for a nominal system G consists in finding an output feedback controller K which stabilizes all multiplicatively perturbed systems $G_\Delta = (I + \Delta)\, G$, where Δ denotes any arbitrary stable dynamic uncertainty with $\| \Delta \|_\infty \leq \delta$, the *robustness radius* $\delta > 0$, being given.

Remark 6.1 If the nominal system G is stable, then the uncertain system G_Δ is stable for any stable Δ, and hence no compensation is necessary to achieve robustness performances with respect to multiplicative uncertainty. □

The robust design problem with respect to multiplicative uncertainty is illustrated in Figure 6.1 from which direct computations give

$$T_{wv} = GK\,(I - GK)^{-1}.$$

Then, based on the Small Gain Theorem (Corollary 2.8), one obtains:

Proposition 6.1 *If K stabilizes G and*

$$\left\| GK(I - GK)^{-1} \right\|_\infty < \gamma := \delta^{-1} \tag{6.1}$$

then K is stabilizing for all systems $G_\Delta = (I + \Delta)G$ with $\Delta \in \mathrm{RH}_+^\infty$ and $\|\Delta\|_\infty \leq \delta$. □

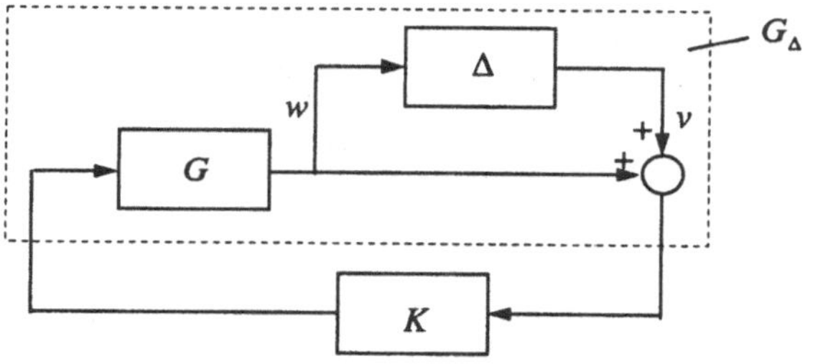

Figure 6.1. **The multiplicative uncertainty modeling**

The problem of robust design with respect to multiplicative uncertainty can be also formulated as an H^∞ problem; indeed, let (A, B, C, D) be a realization of G, where $A \in \mathbf{R}^{n\times n}$, $B \in \mathbf{R}^{n\times m}$, $C \in \mathbf{R}^{p\times n}$, $D \in \mathbf{R}^{p\times m}$ and consider the generalized system

$$T = \begin{bmatrix} T_{11} & T_{12} \\ T_{21} & T_{22} \end{bmatrix} = \left[\begin{array}{c|cc} A & 0 & B \\ \hline C & 0 & D \\ C & I & D \end{array}\right]. \tag{6.2}$$

It is then easy to check that

$$GK(I - GK)^{-1} = LFT(T, K),$$

where LFT denotes the linear fractional transformation defined in Section 3.1.

Therefore, a solution of the robust design problem under investigation may be determined by solving the suboptimal H^∞-control problem associated with T defined by (6.2).

In the case when D is a full-rank column matrix, A is dichotomous and G has no transmission zeros on the $j\mathbf{R}$-axis, the assumptions for the non-singular H^∞ -control problem hold and its solution may be determined by using the formulae derived in Section 6.1. Consider now the case where D is not a full-rank column; this is a typical situation for the strictly proper nominal systems for which $D = 0$. Note that the same situation also appears in design problems when a strictly proper dynamic weighting function W is introduced in order to achieve robustness performances on a certain imposed range of frequency; in this case the attenuation condition becomes

$$\left\| WGK(I - GK)^{-1} \right\|_\infty < \gamma,$$

and the corresponding associated generalized system will have $D_{12} = 0$ if W is strictly proper.

It follows that if the matrix $D_{12}^T D_{12}$ is not invertible the robust design problem with respect to multiplicative uncertainty leads to a singular H^∞ problem.

In order to simplify the presentation for the following developments, we shall consider the singular case when $D = 0$.

6.1.1. NECESSARY AND SUFFICIENT CONDITIONS FOR ROBUST STABILIZATION

Consider the generalized system (6.2) associated with the robust design problem with respect to multiplicative uncertainty. Then according to Theorem 3.14 in Section 3.7, the H^∞-control problem corresponding to this system has a solution if and only if there exist two symmetric matrices $R > 0$ and $S > 0$ such that

$$\begin{bmatrix} \mathcal{N}_{12} & 0 \\ 0 & I \end{bmatrix}^T \begin{bmatrix} AR + RA^T & RC^T & \vdots & 0 \\ CR & -\gamma I & \vdots & 0 \\ \cdots & \cdots & \cdots & \cdots \\ 0 & 0 & \vdots & -\gamma I \end{bmatrix} \begin{bmatrix} \mathcal{N}_{12} & 0 \\ 0 & I \end{bmatrix} < 0, \quad (6.3)$$

$$\begin{bmatrix} \mathcal{N}_{21} & 0 \\ 0 & I \end{bmatrix}^T \begin{bmatrix} A^T S + SA & 0 & \vdots & C^T \\ 0 & -\gamma I & \vdots & 0 \\ \cdots & \cdots & \cdots & \cdots \\ C & 0 & \vdots & -\gamma I \end{bmatrix} \begin{bmatrix} \mathcal{N}_{21} & 0 \\ 0 & I \end{bmatrix} < 0, \quad (6.4)$$

$$\lambda_{\min}(RS) > 1, \quad (6.5)$$

where $\mathcal{N}_{12}$ and $\mathcal{N}_{21}$ are bases of the null spaces of $\begin{bmatrix} B^T & 0_{m\times p} \end{bmatrix}$ and $\begin{bmatrix} C & I_p \end{bmatrix}$, respectively.

The first inequality above reduces to

$$\mathcal{N}_{12}^T \begin{bmatrix} AR + RA^T & RC^T \\ CR & -\gamma I \end{bmatrix} \mathcal{N}_{12} < 0,$$

from which we deduce, using the Finsler's Lemma (Proposition 3.4), that the LMI (6.3) has a solution $R > 0$ if and only if there exists $\alpha > 0$ such that

$$AR + RA^T + \gamma^{-1} RC^T CR - \alpha BB^T < 0.$$

Denoting

$$X := \gamma R^{-1}$$

from the inequality above one obtains

$$A^T X + XA - \alpha\gamma^{-1} XBB^T X + C^T C < 0. \tag{6.6}$$

As concerns condition (6.4), it may be written in a more explicit form by taking into account that

$$\mathcal{N}_{21} = \begin{bmatrix} I \\ -C \end{bmatrix},$$

which gives, after some direct computations, the following inequality equivalent to (6.4)

$$A^T S + SA - \left(\gamma - \gamma^{-1}\right) C^T C < 0.$$

If we denote

$$Y := \gamma S^{-1}$$

the above inequality reduces to

$$AY + YA^T - \left(1 - \gamma^{-2}\right) YC^T CY < 0, \tag{6.7}$$

and with the new variables X and Y the condition (6.5) becomes

$$\rho(XY) < \gamma^2. \tag{6.8}$$

Therefore the following result provides necessary and sufficient conditions for robust stabilization.

Theorem 6.1 *The robust design problem with respect to multiplicative uncertainty has a solution if and only if there exist $\alpha > 0$ and the symmetric matrices $X > 0$ and $Y > 0$ for which the inequalities (6.6), (6.7) and (6.8) hold.* □

A direct consequence of the condition expressed by the Bernoulli inequality (6.7) is the following corollary.

Corollary 6.1 *If the nominal system G with the realization (A, B, C) has a robustness radius $\delta > 1$, then A is stable.*

Proof. Since $\delta > 1$, $1 - \gamma^{-2} = 1 - \delta^2 < 0$, and the inequality (6.7) gives

$$AY + YA^T < 0$$

with $Y > 0$, from which we conclude that A is stable. □

Taking into account the Remark 6.1, the corollary above shows that the case $\delta > 1$ is irrelevant since in this situation no feedback controller is required for robust stabilization. Therefore in the sequel we shall consider only the interesting case $\delta < 1$.

6.1.2. AN EXPLICIT SOLUTION TO THE ROBUST DESIGN PROBLEM

We shall give in this section explicit formulae for the realization of a robust controller with respect to multiplicative uncertainty, in terms of the solutions to the Riccati inequalities (6.6) and (6.7), satisfying the conditions of Theorem 6.1. The main result is given by the following theorem:

Theorem 6.2 *If the conditions in Theorem 6.1 are fulfilled, then a solution to the robust design problem with respect to multiplicative uncertainty is given by the controller K with the realization (A_K, B_K, C_K) with*

$$\begin{aligned} A_K &= A + BC_K - B_K C - Y\left(\gamma^2 I - XY\right)^{-1} M, \\ B_K &= \gamma^2 Y \left(\gamma^2 I - XY\right)^{-1} C^T, \\ C_K &= -\alpha\gamma^{-1} B^T X, \end{aligned} \tag{6.9}$$

where we have denoted

$$M := -A^T X - XA + \alpha\gamma^{-1} XBB^T X - C^T C. \tag{6.10}$$

Proof. When coupling the controller K to the generalized system (6.2) we obtain the resulting system with the realization (A_R, B_R, C_R), where

$$A_R = \begin{bmatrix} A & BC_K \\ B_K C & A_K \end{bmatrix}, \quad B_R = \begin{bmatrix} 0 \\ B_K \end{bmatrix}, \quad C_R = \begin{bmatrix} C & 0 \end{bmatrix}. \tag{6.11}$$

Next, consider the norm Riccati inequality associated with the resulting system, namely

$$\mathcal{P} := A_R^T X_R + X_R A_R + \gamma^{-2} X_R B_R B_R^T X_R + C_R^T C_R < 0. \tag{6.12}$$

In the sequel we shall prove that a solution of this inequality is the positive definite matrix

$$X_R = \begin{bmatrix} X_{R_{11}} & X_{R_{12}} \\ X_{R_{12}}^T & X_{R_{22}} \end{bmatrix} := \begin{bmatrix} \gamma^2 Y^{-1} & -(\gamma^2 Y^{-1} - X) \\ -(\gamma^2 Y^{-1} - X) & \gamma^2 Y^{-1} - X \end{bmatrix} \tag{6.13}$$

where $X > 0$ and $Y > 0$ satisfy the conditions (6.6), (6.7) and (6.8). The fact that X_R is positive definite results from using the nonsingular transformation

$$\widetilde{T} = \begin{bmatrix} I & 0 \\ I & I \end{bmatrix},$$

which gives

$$\widetilde{X}_R = \widetilde{T}^T \widetilde{X}_R \widetilde{T} = \begin{bmatrix} X & 0 \\ 0 & \gamma^2 Y^{-1} - X \end{bmatrix}.$$

Since X and Y are positive-definite and they satisfy (6.8), it follows that $\widetilde{X}_R$ is positive-definite and hence, so is X_R.

We prove now that X_R defined above is a solution of (6.12); to this end consider the nonsingular transformation

$$\widehat{T} := \begin{bmatrix} I & I \\ -I & 0 \end{bmatrix}$$

and the partition of $\mathcal{P}$ defined by (6.12), similar to that of X_R, namely

$$\mathcal{P} = \begin{bmatrix} \mathcal{P}_{11} & \mathcal{P}_{12} \\ \mathcal{P}_{12}^T & \mathcal{P}_{22} \end{bmatrix},$$

for which one obtains

$$\widehat{\mathcal{P}} := \widehat{T}\mathcal{P}\widehat{T}^T = \begin{bmatrix} \mathcal{P}_{11} + \mathcal{P}_{12}^T + \mathcal{P}_{12} + \mathcal{P}_{22} & -\left(\mathcal{P}_{11} + \mathcal{P}_{12}^T\right) \\ -(\mathcal{P}_{11} + \mathcal{P}_{12}) & \mathcal{P}_{11} \end{bmatrix}.$$

We shall show that $\widehat{\mathcal{P}}$ is negative-definite; indeed, when expressing $\mathcal{P}_{11}$ with respect to the elements of X_R by (6.9) and (6.13), it results that

$$\begin{aligned} \mathcal{P}_{11} &= A^T X_{R_{11}} + X_{R_{11}} A + C^T B_K^T X_{R_{12}} + X_{R_{12}} B_K C \\ &\quad + \gamma^{-2} X_{R_{12}} B_K B_K^T X_{R_{12}} + C^T C \\ &= \gamma^2 Y^{-1} \left[YA + A^T Y - (1 - \gamma^{-2}) Y C^T C Y \right] Y^{-1} \\ &< 0. \end{aligned} \tag{6.14}$$

On the other hand, direct algebraic computations give

$$\begin{aligned} \mathcal{P}_{11} + \mathcal{P}_{12}^T &= \gamma^2 Y^{-1} \left[YA + A^T Y - (1 - \gamma^{-2}) Y C^T C Y \right] Y^{-1} \\ &\quad + X_{R_{12}}^T A + X_{R_{22}} B_K C + A_K^T X_{R_{12}}^T + C_K^T B^T X_{R_{11}} \\ &\quad + \gamma^{-2} X_{R_{22}} B_K B_K^T X_{R_{12}}^T \\ &= 0. \end{aligned} \tag{6.15}$$

We also have

$$\begin{aligned} \mathcal{P}_{12} + \mathcal{P}_{22} &= A^T X_{R_{12}} + C^T B_K^T X_{R_{22}} + X_{R_{12}} A_K + X_{R_{11}} B C_K \\ &\quad + \gamma^{-2} X_{R_{12}} B_K B_K^T X_{R_{22}} + A_K^T X_{R_{22}} + X_{R_{22}} A_K + C_K^T B^T X_{R_{12}} \\ &\quad + X_{R_{12}} B C_K + \gamma^{-2} X_{R_{22}} B_K B_K^T X_{R_{22}} \\ &= -M - \alpha\gamma^{-1} X B B^T X. \end{aligned}$$

Since $-M$ is just the left hand side of (6.6), from the equation above it follows that

$$\mathcal{P}_{12} + \mathcal{P}_{22} < 0. \tag{6.16}$$

From (6.14), (6.15) and (6.16) we deduce that $\widehat{\mathcal{P}}$ is a diagonal matrix with negative-definite diagonal elements and therefore, $\widehat{\mathcal{P}} < 0$, from which we conclude that $\mathcal{P} < 0$. Then from the Bounded Real Lemma (Corollary 2.7) it follows that the resulting system (6.11) is stable and its H^∞-norm is less than γ, and hence the controller (6.9) is robust with respect to multiplicative uncertainty, providing a robustness radius $\delta = \gamma^{-1}$. □

Remark 6.2 The singular H^∞ problem resulting when $D = 0$ can be also treated via a singular perturbation technique, by considering instead of (6.2) the augmented generalized system

$$T_\epsilon := \left[\begin{array}{c|ccc} A & 0 & & B \\ \hline C & 0 & \vdots & 0 \\ 0 & 0 & & \varepsilon I \\ \cdots & \cdots & \cdots & \cdots \\ C & I & \vdots & 0 \end{array}\right]$$

with $\epsilon > 0$, defining the input–output dependence

$$\begin{bmatrix} y_1^a \\ y_2 \end{bmatrix} = T_\epsilon \begin{bmatrix} u_1 \\ u_2 \end{bmatrix},$$

where the augmented output vector y_1^a has the partition

$$y_1^a = \begin{bmatrix} y_1 \\ y_{1,\epsilon} \end{bmatrix}.$$

Note that the H^∞ problem corresponding to T_ϵ is nonsingular. Moreover, if K_ϵ is a solution of this problem, then it is a solution of the H^∞ problem associated with the generalized system T; indeed, if $\|LFT(T_\epsilon, K_\epsilon)\|_\infty < \gamma$ it results that for all $u_1 \in \mathrm{L}_+^{2,m}$ we have

$$\|y_1^a\|_2^2 < \gamma^2 \|u_1\|_2^2,$$

and hence there exists $\nu > 0$ small enough such that

$$\begin{aligned} -\nu \|u_1\|_2^2 &\geq \|y_1^a\|_2^2 - \gamma^2 \|u_1\|_2^2 \\ &= \|y_1\|_2^2 + \|y_{1,\epsilon}\|_2^2 - \gamma^2 \|u_1\|_2^2 \\ &> \|y_1\|_2^2 - \gamma^2 \|u_1\|_2^2 \end{aligned}$$

from which we deduce that $\|y_1\|_2^2 < \gamma^2 \|u_1\|_2^2$, that is $\|LFT(T, K_\epsilon)\|_\infty < \gamma$.

Then according to Theorems 3.9 and 3.10, the H^∞-control problem corresponding to T_ϵ has a solution if and only if the AREs

$$\begin{aligned} A^T X + XA - \epsilon^{-2} XBB^T X + C^T C &= 0, \\ AY + YA^T - \left(1 - \gamma^{-2}\right) YC^T CY &= 0, \end{aligned}$$

have stabilizing positive semi-definite solutions X and Y respectively, such that $\rho(XY) < \gamma^2$. If these conditions hold, then a solution of the robust design problem with respect to multiplicative uncertainty is the controller K_ϵ with the realization $(A_{K,\epsilon}, B_{K,\epsilon}, C_{K,\epsilon}, D_{K,\epsilon})$ given by (3.105) updated to the case $\gamma \neq 1$, where

$$\begin{aligned} A_{K,\epsilon} &= A - \epsilon^{-2} BB^T X - ZC^T C, \\ B_{K,\epsilon} &= -ZC^T, \\ C_{K,\epsilon} &= \epsilon^{-2} B^T X, \\ D_{K,\epsilon} &= 0, \quad Z := \gamma^2 Y \left(\gamma^2 I - XY\right)^{-1}. \end{aligned} \tag{6.17}$$

Note that this technique is more restrictive than the one based on LMIs since it requires a dichotomy assumption on A, in order to have a stabilizing solution Y to the Bernoulli equation above. □

6.1.3. STABILITY MARGIN EVALUATION

In Subsection 6.1.1 we proved that if the nominal system G is unstable the robustness radius δ cannot exceed one; recall that the robust stabilization problem with respect to multiplicative uncertainty is irrelevant for stable nominal systems since no compensation is necessary in this case.

In the sequel we shall show that, under some assumptions concerning the unstable nominal system G, one may determine for any $\delta < 1$ a controller providing the robustness radius δ. To this end, firstly we determine some parameterized solutions of the Riccati inequalities (6.6) and (6.7).

As concerns the inequality (6.6), consider its corresponding equation

$$A^T X + XA - \alpha\gamma^{-1} XBB^T X + C^T C = 0. \tag{6.18}$$

Assuming that (A, B, C) is a minimal realization of the nominal system G, it results that the Riccati equation (6.18) has a stabilizing positive-definite solution which we shall denote by $\overline{X}_\alpha$. Since $\overline{X}_\alpha$ is stabilizing it follows that $A - \alpha\gamma^{-1} BB^T \overline{X}_\alpha$ is stable, and therefore the Lyapunov equation

$$\left[A - \alpha\gamma^{-1} BB^T \overline{X}_\alpha\right]^T V + V \left[A - \alpha\gamma^{-1} BB^T \overline{X}_\alpha\right] + I = 0 \tag{6.19}$$

has a unique positive-definite solution V_α. Then define

$$X_{\alpha,\eta} := \overline{X}_\alpha + \eta V_\alpha, \tag{6.20}$$

where $\eta > 0$ is arbitrary; taking into account that $X_{\alpha,\eta}$ and V_α are the solutions of (6.18) and (6.19) respectively, one obtains

$$\begin{aligned} & A^T X_{\alpha,\eta} + X_{\alpha,\eta} A - \alpha\gamma^{-1} X_{\alpha,\eta} BB^T X_{\alpha,\eta} + C^T C \\ = \ & \eta \left[\left(A - \alpha\gamma^{-1} BB^T \overline{X}_\alpha \right)^T V_\alpha + V_\alpha \left(A - \alpha\gamma^{-1} BB^T \overline{X}_\alpha \right) \right] \\ & -\alpha\gamma^{-1}\eta^2 V_\alpha BB^T V_\alpha = -\eta I - \alpha\gamma^{-1}\eta^2 V_\alpha BB^T V_\alpha < 0. \end{aligned}$$

Hence it results that $X_{\alpha,\eta}$ defined by (6.20) is a solution of the Riccati inequality (6.6).

In order to solve the inequality (6.7) consider the equation

$$AY + YA^T - \left(1 - \gamma^{-2}\right) YC^T CY + P = 0 \tag{6.21}$$

where P is an arbitrary symmetric positive-definite matrix with the same dimensions as A. Then since the pair (C, A) is observable and $\left(A, \sqrt{P}\right)$ is controllable it follows that the Riccati equation (6.21) has a unique stabilizing positive definite solution X_P which also verifies (6.7).

Remark 6.3 It is obvious that a similar method to the one used to solve (6.7) might be performed to determine a solution for (6.6), but the solution $X_{\alpha,\eta}$ obtained above is more useful for the further developments. □

In the following, we shall investigate the behavior of the solution $\overline{X}_\alpha$ of (6.18) when α increases; an equivalent problem is obtained if, defining $\epsilon := \sqrt{\alpha^{-1}\gamma}$, we take $\epsilon \searrow 0$. Then (6.18) becomes

$$A^T X + XA - \epsilon^{-2} XBB^T X + C^T C = 0. \tag{6.22}$$

Denote by X_ϵ the stabilizing positive definite solution of (6.22), which exists under the assumptions that the pairs (A, B) and (C, A) are controllable and observable, respectively. Then we have the following:

Proposition 6.2 *If the nominal transfer matrix has a stable right inverse G, then $X_\epsilon \searrow 0$ when $\epsilon \searrow 0$.*

Proof. We shall use some concepts and results from the generalized Popov–Yakubovich theory developed in Chapter 2. Consider the Popov triplet corresponding to (6.22), namely

$$\Sigma = (A, B; Q, L, R),$$

with

$$Q = C^T C, \quad L = 0; \quad R = \epsilon^2 I.$$

For a matrix F such that $A+BF$ is stable (such a matrix exists since the realization (A, B, C) is assumed to be minimal) define the $(0, F)$-equivalent Popov triplet

$$\tilde{\Sigma} := \left(\tilde{A}, \tilde{B}; \tilde{Q}, \tilde{L}, \tilde{R}\right), \tag{6.23}$$

where

$$\begin{aligned}
\tilde{A} &= A + BF, \\
\tilde{B} &= B, \\
\tilde{Q} &= Q + LF + F^T L^T + F^T RF = C^T C + \epsilon^2 F^T F, \\
\tilde{L} &= L + F^T R = \epsilon^2 F^T, \\
\tilde{R} &= R.
\end{aligned}$$

Then according to (3) of Proposition 2.4 the Riccati equation associated with $\tilde{\Sigma}$ has the same stabilizing solution X_ϵ since Σ and $\tilde{\Sigma}$ are $(0, F)$-equivalent. Based on Theorem 2.1, X_ϵ may be also expressed in terms of the operators associated with $\tilde{\Sigma}$, as follows

$$X_\epsilon = \tilde{\Phi}^* \tilde{Q} \tilde{\Phi} - \tilde{\Phi}^* \left(\tilde{Q}\tilde{\mathcal{L}} + \tilde{L}\right) \tilde{\mathcal{R}}^{-1} \left(\tilde{\mathcal{L}}^* \tilde{Q}^T + \tilde{L}^T\right) \tilde{\Phi},$$

where the operators in the right hand side are defined as in Section 2.2 (see (2.27), (2.35)–(2.37)), namely

$$\begin{aligned}
&\tilde{\Phi} : \mathbf{R}^n \to \mathrm{L}_+^{2,n} && \text{for} && \left(\tilde{\Phi}\xi\right)(t) = e^{\tilde{A}t}\xi, \quad t \geq 0, \\
&\tilde{\mathcal{L}} : \mathrm{L}_+^{2,m} \to \mathrm{L}_+^{2,n} && \text{for} && \left(\tilde{\mathcal{L}}u\right)(t) = \int_0^t e^{\tilde{A}(t-\tau)} \tilde{B} u(\tau) d\tau, \quad t \geq 0, \\
&\tilde{\mathcal{R}} : \mathrm{L}_+^{2,m} \to \mathrm{L}_+^{2,m} && \text{for} && \tilde{\mathcal{R}} = \tilde{R} + \tilde{L}^T \tilde{\mathcal{L}} + \tilde{\mathcal{L}}^* \tilde{L} + \tilde{\mathcal{L}}^* \tilde{Q}\, \tilde{\mathcal{L}}.
\end{aligned}$$

Denote by $\tilde{\mathcal{G}}$ the operator

$$\tilde{\mathcal{G}} : \mathrm{L}_+^{2,m} \to \mathrm{L}_+^{2,p} \quad \text{for} \quad \left(\tilde{\mathcal{G}}u\right)(t) = C \int_0^t e^{\tilde{A}(t-\tau)} \tilde{B} u(\tau) d\tau, \ t \geq 0.$$

Since $\tilde{\mathcal{G}} = C\tilde{\mathcal{L}}$, using the definitions above it results that

$$\begin{aligned}
X_\epsilon &= \tilde{\Phi}^* \tilde{Q} \tilde{\Phi} - \tilde{\Phi}^* \left[\left(C^T C + \epsilon^2 F^T F\right) \tilde{\mathcal{L}} + \epsilon^2 F^T\right] \\
&\qquad \times \left[\epsilon^2 I + \epsilon^2 F \tilde{\mathcal{L}} + \epsilon^2 \tilde{\mathcal{L}}^* F^T + \tilde{\mathcal{L}}^* \left(C^T C + \epsilon^2 \tilde{F}^T \tilde{F}\right) \tilde{\mathcal{L}}\right]^{-1} \\
&\qquad \times \left[\tilde{\mathcal{L}}^* \left(C^T C + \epsilon^2 F^T F\right) + \epsilon^2 F\right] \tilde{\Phi} \\
&= \tilde{\Phi}^* \begin{bmatrix} C^T & \epsilon F^T \end{bmatrix} \begin{bmatrix} C \\ \epsilon F \end{bmatrix} \tilde{\Phi} \\
&\quad - \tilde{\Phi}^* \left[C^T \tilde{\mathcal{G}} + \epsilon^2 F^T \left(I + F\tilde{\mathcal{L}}\right)\right] \left[\epsilon^2 \left(I + F\tilde{\mathcal{L}}\right)^* \left(I + F\tilde{\mathcal{L}}\right) + \tilde{\mathcal{G}}^* \tilde{\mathcal{G}}\right]^{-1} \\
&\qquad \times \left[\tilde{\mathcal{G}}^* C + \epsilon^2 \left(I + \tilde{\mathcal{L}}^* \tilde{F}^T\right) F\right] \tilde{\Phi}.
\end{aligned}$$

If we denote

$$\widetilde{H} := I + F\widetilde{\mathcal{L}}, \quad \widetilde{N}_\epsilon := \begin{bmatrix} C \\ \epsilon F \end{bmatrix}, \quad Z_\epsilon := \begin{bmatrix} \widetilde{\mathcal{G}} \\ \epsilon \widetilde{H} \end{bmatrix}$$

then we obtain

$$\begin{aligned} X_\epsilon &= \widetilde{\Phi}^* \widetilde{N}_\epsilon^T \widetilde{N}_\epsilon \widetilde{\Phi} - \widetilde{\Phi}^* \widetilde{N}_\epsilon^T \begin{bmatrix} \widetilde{\mathcal{G}} \\ \epsilon \widetilde{H} \end{bmatrix} \left(\epsilon^2 \widetilde{H}^* \widetilde{H} + \widetilde{\mathcal{G}}^* \widetilde{\mathcal{G}} \right)^{-1} \begin{bmatrix} \widetilde{\mathcal{G}}^* & \epsilon \widetilde{H} \end{bmatrix} \widetilde{N}_\epsilon \widetilde{\Phi} \\ &= \widetilde{\Phi}^* \widetilde{N}_\epsilon^T \left[I - Z_\epsilon \left(Z_\epsilon^* Z_\epsilon \right)^{-1} Z_\epsilon^* \right] \widetilde{N}_\epsilon \widetilde{\Phi}. \end{aligned} \tag{6.24}$$

Define the operator

$$\Pi_\epsilon := I - Z_\epsilon \left(Z_\epsilon^* Z_\epsilon \right)^{-1} Z_\epsilon^*, \tag{6.25}$$

which is a projector; hence

$$\Pi_\epsilon > 0 \tag{6.26}$$

and

$$\|\Pi_\epsilon\| = 1. \tag{6.27}$$

Consider now the partition of Π_ϵ, given by Z_ϵ defined above, namely,

$$\Pi_\epsilon = \begin{bmatrix} \Pi_{\epsilon 11} & \Pi_{\epsilon 12} \\ \Pi_{\epsilon 12}^* & \Pi_{\epsilon 22} \end{bmatrix}, \tag{6.28}$$

where

$$\Pi_{\epsilon 11} = I - \widetilde{\mathcal{G}}^* \left(\epsilon^2 \widetilde{H}^* \widetilde{H} + \widetilde{\mathcal{G}}^* \widetilde{\mathcal{G}} \right)^{-1} \widetilde{\mathcal{G}}. \tag{6.29}$$

Then from (6.26) it results that $\Pi_{\epsilon 11} > 0$, and (6.29) shows that $\Pi_{\epsilon 11}$ decreases when $\epsilon \searrow 0$; therefore the following limit exists:

$$\lim_{\epsilon \searrow 0} \Pi_{\epsilon 11} =: \Pi_{11} \geq 0. \tag{6.30}$$

On the other hand, from the definition of Π_ϵ it follows that

$$Z_\epsilon^* \Pi_\epsilon = 0, \tag{6.31}$$

that is,

$$\begin{bmatrix} \widetilde{\mathcal{G}}^* & \epsilon \widetilde{H}^* \end{bmatrix} \begin{bmatrix} \Pi_{\epsilon 11} & \Pi_{\epsilon 12} \\ \Pi_{\epsilon 12}^* & \Pi_{\epsilon 22} \end{bmatrix} = 0,$$

which reveals that

$$\widetilde{\mathcal{G}}^* \Pi_{\epsilon 11} + \epsilon \widetilde{H}^* \Pi_{\epsilon 12}^* = 0. \tag{6.32}$$

But from (6.27) we deduce that $\|\Pi_{\epsilon 12}\| \leq 1$ for all $\epsilon > 0$; hence (6.32) gives for $\epsilon \searrow 0$,

$$\widetilde{\mathcal{G}}^* \Pi_{11} = 0. \tag{6.33}$$

Since we assumed that G has a stable right inverse one can show that $\tilde{\mathcal{G}}^*$ is one-to-one, and hence (6.33) implies

$$\Pi_{11} = 0. \tag{6.34}$$

Next, write $\tilde{N}_\epsilon$ in the equivalent form

$$\tilde{N}_\epsilon = \begin{bmatrix} I & 0 \\ 0 & \epsilon I \end{bmatrix} \begin{bmatrix} C \\ F \end{bmatrix},$$

and substitute it in (6.24); then, together with (6.25) and (6.28), we obtain

$$\begin{aligned} X_\epsilon &= \tilde{\Phi}^* \begin{bmatrix} C^T & F^T \end{bmatrix} \begin{bmatrix} I & 0 \\ 0 & \epsilon I \end{bmatrix} \begin{bmatrix} \Pi_{\epsilon 11} & \Pi_{\epsilon 12} \\ \Pi^*_{\epsilon 12} & \Pi_{\epsilon 22} \end{bmatrix} \begin{bmatrix} I & 0 \\ 0 & \epsilon I \end{bmatrix} \begin{bmatrix} C \\ F \end{bmatrix} \tilde{\Phi} \\ &= \tilde{\Phi}^* \begin{bmatrix} C^T & F^T \end{bmatrix} \begin{bmatrix} \Pi_{\epsilon 11} & \epsilon \Pi_{\epsilon 12} \\ \epsilon \Pi^*_{\epsilon 12} & \epsilon^2 \Pi_{\epsilon 22} \end{bmatrix} \begin{bmatrix} C \\ F \end{bmatrix} \tilde{\Phi}. \end{aligned} \tag{6.35}$$

Since from (6.27) we have $\|\Pi_{\epsilon 12}\| \leq 1$ and $\|\Pi_{\epsilon 22}\| \leq 1$, it results that $\epsilon \Pi_{\epsilon 12}$ and $\epsilon \Pi_{\epsilon 22}$ tend to zero for ϵ approaching zero. Hence, taking into account (6.34), we conclude that $\lim_{\epsilon \searrow 0} X_\epsilon = 0$, and the proof ends. □

Based on Proposition 6.1, we obtain the main result of this subsection:

Theorem 6.3 *Assume that the nominal system G has a stable right inverse; then for each $0 < \delta < 1$ there exists a controller which stabilizes all systems $G_\Delta = (I + \Delta)G$ with Δ stable and $\| \Delta \|_\infty \leq \delta$; if (A, B, C) is a minimal realization of G, then this controller is given by formulae (6.9) with $\gamma = \delta^{-1}$ and $X = X_{\alpha,\eta}$, $Y = Y_P$, determined as described above, for $\alpha > 0$ large enough and $\eta > 0$ small enough.*

Proof. Let Y_P be a solution of (6.7) where $\gamma = \delta^{-1}$. From Proposition 6.1 it follows that $\overline{X}_\alpha \searrow 0$ for $\alpha \to \infty$; then there exist α large enough and η small enough such that the solution $X_{\alpha,\eta}$ of the Riccati inequality (6.6), given by (6.20), satisfies, together with Y_P, the spectral radius condition (6.8) in which $\gamma = \delta^{-1}$. Hence it results that all conditions in Theorem 6.1 hold and therefore the controller (6.9) with $\gamma = \delta^{-1}$, $X = X_{\alpha,\eta}$, and $Y = Y_P$ is robust with respect to the multiplicative stable uncertainty Δ with $\| \Delta \|_\infty \leq \delta$. □

Theorem 6.3 shows that under the specified assumptions there exists a controller for which explicit formulae are given, providing an imposed robustness radius. If, in addition, the matrix A is dichotomous, the same robustness properties may be obtained using the controller (6.17) derived by a singular perturbation technique; in this case, in order to obtain a robustness radius δ close to 1, it is necessary to take ϵ close to zero.

In the next subsection we shall illustrate by some numerical examples the theoretical results derived above.

6.1.4. NUMERICAL EXAMPLES

To illustrate the results derived in the previous subsections we consider two robust design examples with respect to multiplicative uncertainty. Both of them use strictly proper nominal systems, and hence the design problems reduce to singular H^∞-control problems. In the first example the nominal system has the state matrix A nondichotomous and the robust design problem is solved by using the LMIs-based technique. This example demonstrates that this method can be readily implemented by using standard numerical procedures available in the control toolboxes. In the second design example we considered a nominal system with A dichotomous for which we applied the singular perturbations approach described in Section 6.1.3.

Example 1. Let us consider the nominal system G with the state space matrices given in Section 4.2.3. This system is unstable with A nondichotomous and it has five states, three inputs and three outputs. We computed the solution to the robust design problem with respect to multiplicative uncertainty using the formulae (6.9) and (6.10) in which the LMIs solutions are determined by (6.20) and (6.21), respectively. Different values for δ, α, η and P have been considered, obtaining thus the following results:

Case (1): $\delta = 0.7$, $\alpha = 10$, $\eta = 10^{-1}$, $P = 10^{-2} I_5$.

In this case

$$K = \left[\begin{array}{c|c} A_K & B_K \\ \hline C_K & 0 \end{array}\right],$$

where

$$A_K = \begin{bmatrix} -9.8373 & 0.3050 & -1.3532 & -0.0105 & -0.9774 \\ -0.1753 & -2.8900 & -0.53795 & -0.0580 & 0.5289 \\ -2.4584 & 0.0650 & -0.6710 & 0.9971 & 0.0057 \\ 1.9695 & 0.3496 & -13.8004 & -6.9230 & -0.4672 \\ 1.9262 & -0.0427 & -2.2305 & -0.4350 & -1.8657 \end{bmatrix},$$

$$B_K = \begin{bmatrix} 9.3604 & -0.2905 & 2.3391 \\ -0.2905 & 0.0646 & -0.0619 \\ 2.3391 & -0.0619 & 0.6320 \\ 0.0624 & 0.0084 & 0.0213 \\ 0.0451 & -0.0088 & 0.0137 \end{bmatrix},$$

$$C_K = \begin{bmatrix} 1.3646 & 0.1632 & -1.1662 & -0.8847 & -0.8716 \\ -0.3168 & -2.7487 & -0.5726 & -0.1639 & 0.3541 \\ 2.3994 & 0.2470 & 5.1797 & 1.2958 & -2.6410 \end{bmatrix},$$

which stabilizes G and $\|GK(I - GK)^{-1}\|_\infty = 1.2625\,(\delta = 0.7921)$. The maximal absolute value of all elements of A_K, B_K, C_K is $M_{\max} = 13.8$.

Case (2): $\delta = 0.9999$, $\alpha = 10^3$, $\eta = 10^{-3}$, $P = 10^{-6} \cdot I_5$.

$$A_K = \begin{bmatrix} -0.59561 & 0.0003 & 1.1319 & 0.0000 & -1.0000 \\ -1.1735 & -31.8262 & 0.8836 & 0.0345 & 0.2652 \\ -0.0001 & 0.0000 & 0.0000 & 1.0000 & 0.0000 \\ 42.2265 & 9.7332 & -138.9349 & -17.1960 & -3.6966 \\ 28.1736 & 3.3595 & -33.5603 & -2.5422 & -6.6319 \end{bmatrix},$$

$$B_K = \begin{bmatrix} 0.5950 & -0.0003 & 0.0001 \\ -0.0003 & 0.0000 & 0.0000 \\ 0.0001 & 0.0000 & 0.0000 \\ -0.0001 & 0.0000 & 0.0000 \\ 0.0000 & 0.0000 & 0.0000 \end{bmatrix},$$

$$C_K = \begin{bmatrix} 19.0604 & 2.3354 & -19.8824 & -2.0837 & -4.2210 \\ 1.1135 & -31.4921 & -1.3311 & -0.2155 & -0.3118 \\ 25.2262 & 0.3817 & 30.6753 & 4.2837 & -9.5909 \end{bmatrix},$$

which is stabilizing and $\|GK(I-GK)^{-1}\|_\infty = 1.000000003923\,(\delta = 0.9999)$ and $M_{\max} = 138.9349$. These results demonstrate that the LMIs based procedure allows to determine robust controllers which ensure robustness radii close the optimum, without ill-conditioned computations.

Example 2. Let us consider the nominal system G used in Section 5.1.1 whose state space matrices are given by (5.29). This system is unstable and A is dichotomous and hence it is possible to use the robust design method described in Section 6.1.2 based on the transformation of the singular H^∞ -control problem in a nonsingular H^∞-problem by introducing small perturbations (Remark 6.2). Using formulae (6.17) we determined robust controllers with respect to multiplicative uncertainty for different δ and ϵ, obtaining the following results:

Case (1): $\delta = 0.7$, $\epsilon = 1$

$$A_K = \begin{bmatrix} -8.9769 & 0.9901 & 3.2683 \\ -14.1981 & -1.4524 & -26.5521 \\ 696.4955 & 8.8729 & -272.3735 \end{bmatrix},$$

$$B_K = \begin{bmatrix} -0.1437 & -0.0099 \\ -0.5000 & -0.0344 \\ 0 & 0 \end{bmatrix},$$

$$C_K = \begin{bmatrix} -49.7497 & -0.6338 & 18.4553 \end{bmatrix},$$

which is stabilizing and $\|GK(I - GK)^{-1}\|_\infty = 1.3520\,(\delta = 0.7396)$ and $M_{\max} = 696.4955$.

Case (2): $\delta = 0.986$, $\epsilon = 10^{-2}$

$$A_K = 10^4 \begin{bmatrix} -0.4814 & -0.0006 & 0.1844 \\ -1.6766 & -0.0024 & 0.6392 \\ 7.0119 & 0.0585 & -2.7206 \end{bmatrix},$$

$$B_K = \begin{bmatrix} -94.8827 & -6.5215 \\ -330.7728 & -22.7349 \\ 0 & 0 \end{bmatrix},$$

$$C_K = 10^3 \begin{bmatrix} -5.0085 & -0.0418 & 1.9423 \end{bmatrix},$$

for which the closed-loop system is stable and $\|GK(I - GK)^{-1}\|_\infty = 1.0142\,(\delta = 0.9860)$ and $M_{\max} = 7.0119 \cdot 10^4$.

From the numerical results above it results that the method based on the singular perturbations technique also provides stabilizing controllers ensuring a robustness radius close to the optimum but when approaching this optimum, the computations tend to become ill-conditioned, as $M_{\max}$ indicates.

6.2. SENSITIVITY MINIMIZATION

Given a system G, the sensitivity minimization problem consists in finding a stabilizing controller K such that the H^∞-norm of the *sensitivity function* defined as

$$S := (I - GK)^{-1},$$

is minimized; this problem appears in applications in which the attenuation of the output disturbance to the system output is required. A related problem is the suboptimal case, where a bound $\gamma > 0$ of the sensitivity norm is given and a stabilizing controller K must be determined such that

$$\left\|(I - GK)^{-1}\right\|_\infty < \gamma. \tag{6.36}$$

Since the sensitivity function can be expressed as

$$S = (I - GK)^{-1} = I + GK(I - GK)^{-1},$$

it results that the suboptimal problem (6.36) may be regarded as a H^∞ problem, for which the generalized system has the structure

$$T = \begin{bmatrix} T_{11} & T_{12} \\ T_{21} & T_{22} \end{bmatrix} = \begin{bmatrix} I & G \\ I & G \end{bmatrix}.$$

Let (A, B, C, D) be a realization of G; then a realization of T is given by

$$T := \left[\begin{array}{c|cc} A & 0 & B \\ \hline C & I & D \\ C & I & D \end{array}\right], \tag{6.37}$$

which shows that if $D^T D$ is not invertible, then the H^∞ problem is singular. The same situation may appear in applications when a weighted sensitivity reduction is required, namely

$$\left\| W(I - GK)^{-1} \right\|_\infty < \gamma,$$

where the stable weighting function W is introduced to ensure attenuation properties on some specified frequency range; indeed, if W is strictly proper, one also obtains a singular H^∞ problem.

In the following subsections we consider the singular case when $D = 0$, for which, based on an LMIs technique we shall deduce necessary and sufficient conditions to solve the sensitivity γ-attenuation problem and we shall derive explicit formulae for a solution of this problem. Finally, numerical results illustrate the proposed design methods.

6.2.1. NECESSARY AND SUFFICIENT CONDITIONS

According to Theorem 3.14 in Section 3.7, the H^∞ problem corresponding to the generalized system (6.37) with $D = 0$ has a solution if and only if there exist symmetric positive-definite matrices R and S, satisfying the following conditions

$$\begin{bmatrix} \mathcal{N}_{12} & 0 \\ 0 & I \end{bmatrix}^T \left[\begin{array}{cc:c} AR + RA^T & RC^T & 0 \\ CR & -\gamma I & I \\ \hdashline 0 & I & -\gamma I \end{array}\right] \begin{bmatrix} \mathcal{N}_{12} & 0 \\ 0 & I \end{bmatrix} < 0, \quad (6.38)$$

$$\begin{bmatrix} \mathcal{N}_{21} & 0 \\ 0 & I \end{bmatrix}^T \left[\begin{array}{cc:c} A^T S + SA & 0 & C^T \\ 0 & -\gamma I & I \\ \hdashline C & I & -\gamma I \end{array}\right] \begin{bmatrix} \mathcal{N}_{21} & 0 \\ 0 & I \end{bmatrix} < 0, \quad (6.39)$$

$$\lambda_{\min}(RS) > 1, \quad (6.40)$$

where $\mathcal{N}_{12}$ and $\mathcal{N}_{21}$ are bases of the null spaces of $\begin{bmatrix} B^T & 0 \end{bmatrix}$ and $\begin{bmatrix} C & I \end{bmatrix}$, respectively. Note that, by a Schur complement argument, (6.38) may be written in the equivalent form

$$\mathcal{N}_{12}^T \begin{bmatrix} AR + RA^T & RC^T \\ CR & -(\gamma - \gamma^{-1}) I \end{bmatrix} \mathcal{N}_{12} < 0,$$

from which we deduce, using Finsler's lemma (Proposition 3.4), that (6.38) has a symmetric positive definite solution R if and only if there exists $\alpha > 0$

such that

$$\begin{bmatrix} AR + RA^T - \alpha BB^T & RC^T \\ CR & -\gamma\left(1-\gamma^{-2}\right)I \end{bmatrix} < 0.$$

The inequality above leads, by using again the Schur complement formula, to the conditions

$$\gamma > 1 \tag{6.41}$$

and

$$AR + RA^T + \frac{\gamma}{\gamma^2 - 1} RC^T CR - \alpha BB^T < 0. \tag{6.42}$$

If we denote

$$X := \gamma R^{-1}$$

then (6.42) reduces, after pre- and post-multiplication by R^{-1}, to

$$A^T X + XA - \alpha\gamma^{-1} XBB^T X + \frac{\gamma^2}{\gamma^2 - 1} C^T C < 0. \tag{6.43}$$

Next, write the condition (6.39) in a more explicit form; taking into account that

$$\mathcal{N}_{21} = \begin{bmatrix} I \\ -C \end{bmatrix}$$

is a basis of the null space of $\begin{bmatrix} C & I \end{bmatrix}$, (6.39) gives

$$A^T S + SA - \gamma C^T C < 0, \tag{6.44}$$

from which, by denoting

$$Y := \gamma S^{-1},$$

one obtains

$$AY + YA^T - YC^T CY < 0. \tag{6.45}$$

Finally, the condition (6.40) can be expressed in terms of X and Y, defined above, as

$$\rho(XY) < \gamma^2. \tag{6.46}$$

The developments above directly lead to the following theorem, which provides necessary and sufficient conditions for solving the suboptimal sensitivity problem.

Theorem 6.4 *Given the realization (A, B, C) of the plant G, the sensitivity γ-attenuation problem with $\gamma > 1$ has a solution if and only if there exist a scalar $\alpha > 0$ and two symmetric positive-definite matrices X and Y, satisfying the conditions (6.43), (6.45) and (6.46).* □

6.2.2. EXPLICIT FORMULAE FOR THE CONTROLLER

In this subsection we shall give explicit formulae for the solution of the sensitivity attenuation problem, under the assumption that the conditions in Theorem 6.4 are fulfilled. These formulae are provided by the following result:

Theorem 6.5 *Assume that there exist symmetric positive-definite matrices X and Y and a scalar $\alpha > 0$ such that (6.43), (6.45) and (6.46) with $\gamma > 1$ hold. Then a solution of the sensitivity attenuation problem $\|(I - GK)^{-1}\|_\infty < \gamma$ is $K = \left[\begin{array}{c|c} A_K & B_K \\ \hline C_K & 0 \end{array}\right]$, where*

$$\begin{aligned} A_K &= A - \tfrac{\alpha}{2\gamma} BB^T X - Y\left(\gamma^2 I - XY\right)^{-1}\left(M + \gamma^2 C^T C\right), \\ B_K &= \gamma^2 Y\left(\gamma^2 I - XY\right)^{-1} C^T, \\ C_K &= -\tfrac{\alpha}{2\gamma} B^T X, \\ M &:= -\left(A^T X + XA - \tfrac{\alpha}{2\gamma} XBB^T X\right). \end{aligned} \tag{6.47}$$

Proof. The same technique used to prove Theorem 6.2 may be applied in this case, namely, it is shown that there exists a positive-definite matrix X_R satisfying the norm type of Riccati inequality corresponding to the resulting system (A_R, B_R, C_R, D_R), which result ensures, according to Bounded Real Lemma (Corollary 2.7), that K is stabilizing and a γ-attenuator. In this case the Riccati inequality is

$$\begin{aligned} \mathcal{P} \;:=\; & A_R^T X_R + X_R A_R + \left(\gamma^2 - 1\right)^{-1}\left(X_R B_R + C_R^T D_R\right) \\ & \times\left(B_R^T X_R + D_R^T C_R\right) + C_R^T C_R < 0, \end{aligned} \tag{6.48}$$

and the matrices A_R, B_R, C_R and D_R have the following expressions

$$\begin{aligned} A_R &= \begin{bmatrix} A & BC_K \\ B_K C & A_K \end{bmatrix}, & B_R &= \begin{bmatrix} 0 \\ B_K \end{bmatrix}, \\ C_R &= \begin{bmatrix} C & 0 \end{bmatrix}, & D_R &= I. \end{aligned}$$

Based on an identical argument to that in the proof of Theorem 6.2, it is shown that

$$X_R = \begin{bmatrix} \gamma^2 Y^{-1} & -\left(\gamma^2 Y^{-1} - X\right) \\ -\left(\gamma^2 Y^{-1} - X\right) & \gamma^2 Y^{-1} - X \end{bmatrix}$$

is positive-definite; by direct algebraic computations, one can also prove, using the same steps as in the proof of Theorem 6.2, that X_R is a solution of (6.48). □

If (A, B, C) is a minimal realization of G, then the inequalities (6.43) and (6.45) can be solved by considering the corresponding Riccati equations

$$A^T X + XA - \alpha\gamma^{-1} XBB^T X + \frac{\gamma^2}{\gamma^2 - 1} C^T C + P = 0 \tag{6.49}$$

and

$$AY + YA^T - YC^T CY + P = 0 \tag{6.50}$$

respectively, where P is an arbitrary symmetric positive-definite matrix; these equations have the stabilizing positive definite solutions $X_{\alpha,P}$ and Y_P, respectively, and obviously they satisfy (6.43) and (6.45), respectively.

Remark 6.4 If G has a stable right inverse then, according to Proposition 6.2, $X_{\alpha,P} \searrow 0$ for $\alpha \to \infty$, and hence for every $\gamma > 1$ there exists an $\alpha > 0$ large enough such that the spectral norm condition $\rho(X_{\alpha,P} Y_P) < \gamma$ is satisfied, and this level of attenuation can be obtained by the controller (6.47) with $X = X_{\alpha,P}$ and $Y = Y_P$. □

6.2.3. THE WEIGHTED SENSITIVITY MINIMIZATION

In many applications when the frequency bandwidth of the disturbance is assumed known, a weighted sensitivity is considered, namely

$$S = W(I - GK)^{-1},$$

where W denotes a stable weighting function chosen such that its magnitude is large enough on the frequency domain of interest and appropriately small in the rest. The weighted sensitivity may be expressed as

$$S = W + WGK(I - GK)^{-1},$$

which shows that the attenuation problem is equivalent to an H^∞-problem for the generalized system

$$T = \begin{bmatrix} T_{11} & T_{12} \\ T_{21} & T_{22} \end{bmatrix} = \begin{bmatrix} W & WG \\ I & G \end{bmatrix}.$$

Let (A_w, B_w, C_w, D_w) be a realization of W, where $A_w \in \mathbf{R}^{w \times w}$ is stable, and assume again that G is strictly proper, namely $D = 0$; then a realization of the generalized system above is

$$\left(\hat{A}, \begin{bmatrix} \hat{B}_1 & \hat{B}_2 \end{bmatrix}, \begin{bmatrix} \hat{C}_1 \\ \hat{C}_2 \end{bmatrix}, \begin{bmatrix} D_w & 0 \\ I & 0 \end{bmatrix}\right), \tag{6.51}$$

where

$$\begin{aligned}\widehat{A} &= \begin{bmatrix} A & 0 \\ B_wC & A_w \end{bmatrix}; \quad \widehat{B}_1 = \begin{bmatrix} 0 \\ B_w \end{bmatrix}; \quad \widehat{B}_2 = \begin{bmatrix} B \\ 0 \end{bmatrix}; \\ \widehat{C}_1 &= \begin{bmatrix} D_wC & C_w \end{bmatrix}; \quad \widehat{C}_2 = \begin{bmatrix} C & 0 \end{bmatrix}. \end{aligned} \qquad (6.52)$$

Since $D_{12} = 0$, we have again obtained a singular H^∞-problem which may be treated in a similar way to the unweighted sensitivity case. Indeed, the necessary and sufficient conditions given by Theorem 3.14 are expressed as

$$\begin{bmatrix} \widehat{\mathcal{N}}_{12} & 0 \\ 0 & I \end{bmatrix}^T \begin{bmatrix} \widehat{A}R + R\widehat{A}^T & R\widehat{C}_1^T & \vdots & \widehat{B}_1 \\ \widehat{C}_1R & -\gamma I & \vdots & D_w \\ \cdots & \cdots & \cdots & \cdots \\ \widehat{B}_1^T & D_w^T & \vdots & -\gamma I \end{bmatrix} \begin{bmatrix} \widehat{\mathcal{N}}_{12} & 0 \\ 0 & I \end{bmatrix} < 0, \quad (6.53)$$

$$\begin{bmatrix} \widehat{\mathcal{N}}_{21} & 0 \\ 0 & I \end{bmatrix}^T \begin{bmatrix} \widehat{A}^TS + S\widehat{A} & S\widehat{B}_1 & \vdots & \widehat{C}_1^T \\ \widehat{B}_1^TS & -\gamma I & \vdots & D_w^T \\ \cdots & \cdots & \cdots & \cdots \\ \widehat{C}_1 & D_w & \vdots & -\gamma I \end{bmatrix} \begin{bmatrix} \widehat{\mathcal{N}}_{21} & 0 \\ 0 & I \end{bmatrix} < 0, \quad (6.54)$$

$$\lambda_{\min}(RS) > 1, \qquad (6.55)$$

where $\widehat{\mathcal{N}}_{12}$ and $\widehat{\mathcal{N}}_{21}$ are bases of the null spaces of $\begin{bmatrix} \widehat{B}_2^T & 0 \end{bmatrix}$ and $\begin{bmatrix} \widehat{C}_2 & I \end{bmatrix}$, respectively.

The inequality (6.53) is equivalent to

$$\widehat{\mathcal{N}}_{12}^T \begin{bmatrix} \widehat{A}R + R\widehat{A}^T + \gamma^{-1}\widehat{B}_1\widehat{B}_1^T & R\widehat{C}_1^T + \gamma^{-1}\widehat{B}_1D_w^T \\ \widehat{C}_1R + \gamma^{-1}D_w\widehat{B}_1^T & -\gamma I + \gamma^{-1}D_wD_w^T \end{bmatrix} \widehat{\mathcal{N}}_{12} < 0,$$

which, according to the Finsler's lemma (Proposition 3.4), holds if and only if that there exits $\alpha > 0$ such that

$$\begin{bmatrix} \widehat{A}R + R\widehat{A}^T + \gamma^{-1}\widehat{B}_1\widehat{B}_1^T - \alpha\widehat{B}_2\widehat{B}_2^T & R\widehat{C}_1^T + \gamma^{-1}\widehat{B}_1D_w^T \\ \widehat{C}_1R + \gamma^{-1}D_w\widehat{B}_1^T & -\gamma I + \gamma^{-1}D_wD_w^T \end{bmatrix} < 0.$$

From the inequality above we deduce by a Schur complement argument the following conditions

$$\gamma^2 I - D_wD_w^T > 0, \qquad (6.56)$$

and

$$\begin{aligned} &\widehat{A}R + R\widehat{A}^T + \gamma^{-1}\widehat{B}_1\widehat{B}_1^T - \alpha\widehat{B}_2\widehat{B}_2^T \\ &\quad + \gamma^{-1}\left(\gamma R\widehat{C}_1^T + \widehat{B}_1D_w^T\right)\left(\gamma^2 I - D_wD_w^T\right)^{-1}\left(\gamma\widehat{C}_1R + D_w\widehat{B}_1^T\right) < 0. \end{aligned}$$

Denoting $X := \gamma R^{-1}$, the last inequality may be rewritten, after pre- and post-multiplying by γR^{-1}, in the equivalent form

$$\begin{aligned} \widetilde{A}^T X + X\widetilde{A} + X\left[\widehat{B}_1\left(\gamma^2 I - D_w^T D_w\right)^{-1}\widehat{B}_1^T - \alpha\gamma^{-1}\widehat{B}_2\widehat{B}_2^T\right]X \\ +\gamma^2\widehat{C}_1^T\left(\gamma^2 I - D_w D_w^T\right)^{-1}\widehat{C}_1 < 0 \end{aligned} \quad (6.57)$$

where

$$\widetilde{A} := \widehat{A} + \widehat{B}_1 D_w^T\left(\gamma^2 I - D_w D_w^T\right)^{-1}\widehat{C}_1. \quad (6.58)$$

To write (6.54) in a more explicit form we take into account that

$$\widehat{\mathcal{N}}_{21} = \begin{bmatrix} I \\ -\widehat{C}_2 \end{bmatrix},$$

which gives, after some direct computations, the condition, equivalent to (6.54),

$$\overline{A}Y + Y\overline{A}^T + Y\overline{Q}Y < 0, \quad (6.59)$$

where we have denoted

$$\begin{aligned} Y &:= \gamma S^{-1}, \\ \overline{A} &:= \begin{bmatrix} A & 0 \\ 0 & A_w \end{bmatrix}, \\ \overline{Q} &:= \begin{bmatrix} -C^T C & 0 \\ 0 & \gamma^{-2}C_w^T C_w \end{bmatrix}. \end{aligned} \quad (6.60)$$

Then the following theorem, providing necessary and sufficient conditions to solve the weighted sensitivity attenuation problem, is obtained:

Theorem 6.6 *The weighted sensitivity attenuation problem*

$$\left\|W(I - GK)^{-1}\right\|_\infty < \gamma$$

has a solution if and only if:

i) Condition (6.56) holds;

ii) There exist $\alpha > 0$ and the positive definite matrices X and Y satisfying the conditions (6.57), (6.59) and $\rho(XY) < \gamma^2$. □

Remark 6.5 From the structure of $\overline{A}$ and $\overline{Q}$ defined by (6.60) it results that a solution Y to (6.59) is given by $Y = \operatorname{diag}(Y_1, Y_2)$, where Y_1 and Y_2 are positive definite matrices such that

$$AY_1 + Y_1 A^T - C^T C < 0 \quad (6.61)$$

and

$$A_w Y_2 + Y_2 A_w^T + \gamma^{-2} C_w^T C_w < 0, \tag{6.62}$$

respectively. Note that if the pair (C, A) is detectable then (6.61) has positive-definite solutions and (6.62) also has positive-definite solutions, since we have assumed that A_w is stable. □

An explicit solution of the weighted sensitivity attenuation problem is provided by the following result:

Theorem 6.7 *Assume that the condition (6.56) holds and that there exist* $\alpha > 0$ *and matrices* X, Y_1, Y_2 *satisfying (6.57), (6.61) and (6.62) respectively, such that* $\rho(XY) < \gamma^2$, *where* $Y = \operatorname{diag}(Y_1, Y_2)$; *then a solution to the weighted* γ*-attenuation problem is given by* $K = \left[\begin{array}{c|c} A_K & B_K \\ \hline C_K & 0 \end{array}\right]$ *with*

$$\begin{aligned} A_K &= \widehat{A} - \tfrac{\alpha}{2\gamma}\widehat{B}_2\widehat{B}_2^T X - \widehat{Y}\left(\gamma^2 I - X\widehat{Y}\right)^{-1}\left(M - \gamma^2\widehat{Y}^{-1}L\widehat{C}_2\right), \\ B_K &= \gamma^2\left(\gamma^2 I - \widehat{Y}X\right)^{-1} L, \\ C_K &= -\tfrac{\alpha}{2\gamma}\widehat{B}_2^T X, \end{aligned} \tag{6.63}$$

where we have denoted

$$\begin{aligned} \widehat{Y} &:= \begin{bmatrix} I_n & 0 \\ 0 & \gamma^{-2}\eta^{-1} I_w \end{bmatrix} Y, \\ L &:= \begin{bmatrix} Y_1 C^T \\ B_w \end{bmatrix}, \\ M &:= -\frac{\alpha}{2\gamma} X\widehat{B}_2\widehat{B}_2^T X + \widehat{Y}\left(\gamma^2 I - X\widehat{Y}\right)^{-1}\left(B_K - \widehat{B}_1\right)\left(\gamma^2 I - D_w^T D_w\right)^{-1} \\ &\qquad \times \left(\widehat{B}_1^T X + D_w^T \widehat{C}_1\right) + N, \qquad (6.64) \\ N &:= -\left[\left(\widehat{A} - \frac{\alpha}{2\gamma}\widehat{B}_2\widehat{B}_2^T X\right)^T X + X\left(\widehat{A} - \frac{\alpha}{2\gamma}\widehat{B}_2\widehat{B}_2^T X\right)\right. \\ &\qquad \left. + \left(X\widehat{B}_1 + \widehat{C}_1^T D_w\right)\left(\gamma^2 I - D_w^T D_w\right)^{-1}\left(\widehat{B}_1^T X + D_w^T \widehat{C}_1\right) + \widehat{C}_1^T \widehat{C}_1\right] \\ \eta &:= \lambda_{\min}^{-1}\left(\gamma^2 I - D_w D_w^T\right), \end{aligned}$$

and the matrices $\widehat{A}$, $\widehat{B}_1$, $\widehat{B}_2$, $\widehat{C}_1$, $\widehat{C}_2$ *are given by (6.52).*

Proof. The proof is also based on the Bounded Real Lemma (Corollary 2.7) and it consists in showing that the norm type of Riccati inequality associated with the resulting system $W(I - GK)^{-1}$ has a positive definite solution; this solution is

$$X_R = \begin{bmatrix} \gamma^2\widehat{Y}^{-1} & -\left(\gamma^2\widehat{Y}^{-1} - X\right) \\ -\left(\gamma^2\widehat{Y}^{-1} - X\right) & \gamma^2\widehat{Y}^{-1} - X \end{bmatrix}.$$

Note that an argument similar to that for Theorem 6.2 can be used to show that X_R is positive-definite; additionally, in this case the condition $\rho\left(X\widehat{Y}\right) < \gamma^2$ is required, but based on the definitions of $\widehat{Y}$ and η it is easy to show that this condition holds since $\rho(XY) < \gamma^2$.

The proof that X_R satisfies the norm type of Riccati inequality corresponding to the system $W(I - GK)^{-1}$ is similar to that in Theorem 6.5 and therefore it is omitted. □

6.2.4. NUMERICAL EXAMPLES

In order to illustrate the results derived in the previous subsection we consider the longitudinal dynamics G of the aircraft which state space realization is given in Section 4.2.3, and we solve the unweighted sensitivity and the weighted sensitivity attenuation problems for this system.

The unweighted sensitivity case. Applying the formulae (6.47) in which the solutions X and Y of the Riccati inequalities (6.43) and (6.45) are determined by solving (6.49) and (6.50), we obtained for $\gamma = 1.1$, $\alpha = 10^5$ and $P = I_5$ we obtained the controller K with the realization (A_K, B_K, C_K), where

$$A_K = \begin{bmatrix} -4.6936 & 0.3615 & 2.3531 & 0.5345 & -2.6494 \\ -0.6086 & -39.4442 & 2.2823 & 0.9728 & 0.7446 \\ -0.6719 & 0.0470 & 0.1534 & 1.0874 & -0.2436 \\ 18.9650 & 6.8390 & -176.1911 & -69.5425 & -22.1694 \\ 24.9387 & 2.5291 & -44.4689 & -18.4114 & -15.4781 \end{bmatrix},$$

$$B_K = \begin{bmatrix} 1.3016 & -0.0872 & 0.1895 \\ -0.0872 & 0.0122 & -0.0112 \\ 0.1895 & -0.0112 & 0.0359 \\ -0.0113 & 0.0020 & -0.0009 \\ 0.0127 & -0.0017 & 0.0021 \end{bmatrix},$$

$$C_K = \begin{bmatrix} 17.4932 & 1.8211 & -26.6049 & -11.9145 & -10.0273 \\ 1.1725 & -39.1235 & -0.6545 & -0.4225 & -0.6393 \\ 35.0629 & 0.7502 & 35.2026 & 9.6292 & -13.8982 \end{bmatrix}$$

for which the eigenvalues of $(I - GK)^{-1}$ are:

$$\begin{gathered} -74.0235 \\ -39.1883 \\ -6.2903 \pm 6.8322i \\ -2.9032 \\ -0.7802 \pm 1.0298i \end{gathered}$$

$$-0.1134 \pm 0.2263i$$
$$-0.1174,$$

and $\|(I - GK)^{-1}\|_\infty = 1.0472.$

The weighted sensitivity case. Assume that the design objective is the sensitivity attenuation for low frequency disturbances at $\omega < 0.1$rad/s. Then we chose the weighting function $W(s) = (10s + 100)/(10s + 1)$ which has a large amplitude on the frequency domain under interest and we determined a solution to the weighted sensitivity attenuation problem using formulae (6.63) and (6.64). For $\gamma = 25$, $\alpha = 10^5$ we obtained the controller

$$K = \left[\begin{array}{c|c} A_K & B_K \\ \hline C_K & 0 \end{array}\right],$$

where

$$A_K = [\ A_{K_1} \quad A_{K_2}\ ],$$

$$A_{K_1} = \begin{bmatrix} -0.3801 & 0.0218 & 1.0656 & 0.0002 \\ -0.9876 & -14.2122 & 0.4055 & 0.0139 \\ -0.0666 & 0.0007 & -0.0261 & 1.0000 \\ 27.6927 & 3.0799 & -102.5967 & -10.5280 \\ 29.7985 & 0.8877 & -20.7615 & -0.6299 \\ -0.0201 & 0.0006 & 0.0012 & 0.0014 \\ 0.0002 & 0.0036 & -0.0002 & 0.0000 \\ -0.0059 & 0.0000 & -0.0021 & 0.0017 \end{bmatrix},$$

$$A_{K_2} = \begin{bmatrix} -1.0007 & -0.0126 & 0.0003 & -0.0032 \\ 0.1598 & -3.4315 & -99.3141 & 4.0815 \\ -0.0001 & -0.0022 & 0.0000 & -0.0008 \\ -2.2498 & 132.1612 & 29.1577 & -443.1923 \\ -5.4855 & 87.2519 & 10.2491 & -127.2117 \\ -0.0055 & -0.0413 & 0.0000 & -0.0007 \\ 0.0001 & 0.0000 & -0.0386 & 0.0000 \\ -0.0015 & -0.0007 & 0.0000 & -0.0394 \end{bmatrix},$$

$$B_K = \begin{bmatrix} 0.3801 & -0.0217 & 0.0673 \\ -0.0220 & 0.0160 & -0.0008 \\ 0.0667 & -0.0007 & 0.0263 \\ -0.0020 & 0.0029 & 0.0012 \\ 0.0032 & -0.0021 & 0.0012 \\ 1.0468 & -0.0005 & 0.0095 \\ -0.0003 & 1.0051 & 0.0002 \\ 0.0010 & 0.0002 & 1.0194 \end{bmatrix},$$

$$C_K = [\ C_{K_1} \quad C_{K_2}\ ],$$

$$C_{K_1} = \begin{bmatrix} 20.7025 & 0.7554 & -11.7690 & -0.9110 \\ 1.4747 & -14.0518 & -0.8364 & -0.0954 \\ 38.3145 & 0.1826 & 30.3832 & 3.3914 \end{bmatrix},$$

$$C_{K_2} = \begin{bmatrix} -3.4368 & 58.9848 & 6.4946 & -78.0222 \\ -0.3232 & 3.6459 & -58.5346 & -5.2813 \\ -8.3786 & 77.1729 & -0.2752 & 59.1063 \end{bmatrix},$$

for which the poles of the closed loop system are:

$$\begin{array}{l} -2.6582 \pm 8.7276i \\ -7.1193 \pm 7.0812i \\ -5.9428 \\ -0.9321 \pm 3.8771i \\ -2.9740 \\ -0.7803 \pm 1.0300i \\ -0.1543 \pm 0.2506i \\ -0.1413, \end{array}$$

and the diagram of the maximal singular value of $(I - GK)^{-1}$ is shown in Figure 6.2, indicating a significant attenuation of the sensitivity at $\omega < 0.1$rad/s.

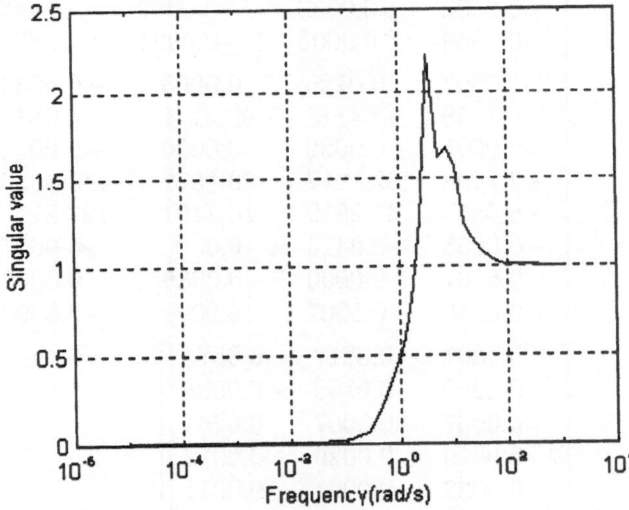

Figure 6.2. **The maximal singular value of $(I - GK)^{-1}$**

From our computational experience we have found that some erroneous results can be obtained in applications in which the numerical procedures do not provide correct solutions to the Riccati inequalities (6.57) and (6.59).

NOTES AND REFERENCES

Early references concerning the robust design problem with respect to multiplicative uncertainty include (Doyle and Stein, 1981) and (Chen and Desoer, 1982) in which robust stability conditions are derived.

In (Vidyasagar and Kimura, 1986) the problem is solved under some assumptions, by reducing it to a one-block Nehari problem using the parameterization of all stabilizing controllers (Youla *et al.*, 1976). In (McFarlane and Glover, 1990) it is shown that the problem can be transformed into an H^∞-control problem for an appropriate generalized system. The singular H^∞-control problem arising in the robust design with respect to multiplicative uncertainty has been addressed in (Stoorvogel, 1991), and more recently in (Dragan *et al.*, 1996) in which the LMIs-based technique described in Section 6.1 has been used.

The sensitivity minimization problem originates in (Zames, 1981). Different methods have been proposed for solving this problem. For linear time-invariant systems we mention the ones given in (Francis and Zames, 1984), (Chang and Pearson, 1984), (Francis *et al.*, 1984), (Safonov and Verma, 1985) in which different H^∞ minimization approaches based on interpolation techniques and operator theory have been used. A similar result with the one stated in Remark 6.4 can be found in (Francis, 1986). Further developments in which the weighted sensitivity problem is addressed, together with other design specifications, can be found, for instance, in (McFarlane and Glover, 1990).

BIBLIOGRAPHY

Adamjan, V.M., Arov, D.Z. and Krein, M.G. (1978) Infinite block Hankel matrices and related extension problems, *AMS Transl.*, **Vol. 111**, pp. 133–156.

Anderson, B.D.O. (1967) An algebraic solution to the spectral factorization problem, *IEEE-Trans. Autom. Control*, **Vol. 12**, pp .410–414.

Ball J.A. and Helton, J.W. (1983) A Beurling-Lax theorem for the Lie group $U(m,n)$ which contains most classical interpolation theory, *J. Op. Theory*, **Vol. 9**, pp. 107–142.

Ball, J.A., Gohberg, I. and Rodman, L. (1990) Nehari interpolation problem for rational matrix: The generic case, *H^∞- Control Theory,* Como 1990, Springer.

Bart, H., Gohberg, I. and Kaashoek, M.A. (1979) *Minimal factorization of matrix and operator functions*, Birkhäuser.

Ben-Artzi, A., Gohberg, I. and Kaashoek, M.A. (1995) *Discrete nonstationary bounded real lemma in indefinite matrix; the strict contractive case*, Operator Theory: Advances and Applications, **80**, Birkhäuser.

Boyd, S., El Ghaoui, L., Feron, E. and Balakrishnan, V. (1994) *Linear matrix inequalities in systems and control theory*, Studies in Applied Mathematics, SIAM, Philadelphia, PA, **Vol. 15**.

Cavallo, A., De Maria, G. and Verde, L. (1992) Robust flight control systems: a parameter space design, *Journal of Guidance, Control and Dynamics*, **Vol. 15, no. 5**, pp. 1207–1215.

Chang, B.C. and Pearson, J.B. (1984) Optimal disturbance rejection in linear multivariable systems, *IEEE-Trans. Autom. Control*, **Vol. 29**, pp. 880–887.

Chang, B.C. and Pearsen, J.B. (1985) Iterative computation of minimal H^∞-norm, *Proceedings of the 24th Conference on Decision and Control, Dec. 1985.*

Chang, B.C., Banda, S.S. and McQuade (1989) Fast iterative computation of optimal two-block H^∞-norm, *IEEE-Trans. Autom. Control*, **Vol. 34, no. 7**, pp. 738–743.

Chen, M. and Desoer, C. (1982) Necessary ans sufficient conditions for robust stability of linear distributed feedback systems, *Int. J. Control*, **Vol. 35**, pp. 255–267.

Chu, C.C., Doyle, J. and Lee, E.B. (1986) The general distance problem in H^∞ optimal control theory, *Int. J. Control*, **Vol. 40**, pp. 565–596.

Coppel, W.A. (1975) Linear quadratic optimal control, *Proceedings of the Royal Society, Edinburgh A78*, pp. 271-289.

Doyle, J.C. and Stein, G. (1981) Multivariable feedback design: concepts for a classical/modern synthesis, *IEEE-Trans. Autom. Control*, **Vol. 26**, pp. 4–16.

Doyle, J.C. (1983) Synthesis of robust controllers and filters, *Proceedings of the 22nd IEEE Conference on Decision and Control, Dec. 1983.*

Doyle, J.C., Glover, K., Khargonekar, P. and Francis, P. (1989) State-space solutions to standard H_2 and H_∞ control problem, *IEEE-Trans-Autom. Control*, **Vol. 34, no. 8**, pp. 831–848.

Dragan, V., Halanay, A. and Stoica, A. (1995a) Two-block Nehari and H^∞ problems, *Preprint Series of the Institute of Mathematics of the Romanian Academy*, **No. 1**, pp. 1-19.

Dragan, V., Halanay, A. and Stoica, A. (1995b) A procedure to compute an optimal robust controller, *Proceedings of European Control Conference, Sept. 1995*, pp. 1005–1010.

Dragan, V., Halanay, A. and Stoica, A. (1995c) Remark on order reduction for a robustly suboptimal controller via singular perturbations, *Systems & Control Letters*, **Vol. 24**, pp. 317–320.

Dragan, V., Halanay, A. and Stoica, A. (1996) On robustness with respect to multiplicative perturbations, *Revue Roumaine des Sciences Téchniques, Serie Eléctrotéchnique et Energétique*, **Vol. 41, no. 3**, pp. 363–377.

Dragan, V., Halanay, A. and Stoica, A. (1997) Well conditioned computation for H^∞ controller near the optimum, *Numerical Algorithms*, **Vol. 15**, pp. 193–206.

Dragan, V. and Stoica, A. (1997) Optimal H^∞ design of a glide-slope coupler, *Preprints of the Fourth IFAC Conference on System Structure and Control, October 23-25, 1997, Bucharest, Romania*, pp. 273–277.

Dym, H. and Gohberg, I. (1988) A new class of contractive interpolants and maximum entropy principles, *Topics in operator theory* (I. Gohberg and M.A. Kaashoek, Eds.), Birkhauser, Basel, pp. 117–150.

Francis, B.A. (1983) Notes on H^∞-optimal linear feedback systems, Lectures presented in the Division of Autom. Contr., Dep. Elect. Eng., Linköping Univ., Linköping, Sweden, Aug. 1983.

Francis, B.A. and Zames, G. (1984) On H^∞-optimal sensitivity theory for SISO feedback systems, *IEEE-Trans. Autom. Control*, **Vol. 29**, no. 1, pp. 9–16.

Francis, B.A., Helton, J.W. and Zames, G. (1984) H^∞ optimal feedback controllers for linear multivariable systems, *IEEE-Trans. Autom. Control*, **Vol. 29**, pp. 888–900.

Francis, B.A. (1986) *A course in H_∞ control theory*, Springer Verlag.

Franklin, S.N. and Ackermann, J. (1981) Robust flight control: a design example, *Journal of Guidance and Control*, **Vol. 4, no. 6**, pp. 597–605.

Feintuch, A. and Francis, B.A. (1986) Uniformly optimal control of linear feedback systems, *Automatica*, **Vol. 21, no. 5**, pp. 563–574.

Gahinet, P. (1992a) Reliable computation of H^∞ central controllers near the optimum, *Rapport de Recherche no. 1642, INRIA*.

Gahinet, P. (1992b) On the game Riccati equations arising in H_∞ control problems, *Rapport de Recherche, no. 1643, INRIA*.

Gahinet, P. and Apkarian, P. (1994) A linear matrix inequality approach to H_∞ control, *International Journal Robust Nonlinear Control*, **No. 4**, pp.421–448.

Gahinet, P., Nemirowski, A., Laub, A.J. and Chilali, M. (1995) *LMI Control Toolbox*, The MathWorks Inc.

Glover, K. (1984) All-optimal Hankel-norm approximations of linear multivariable systems and their L_∞-error bounds, *Int. J. Control*, **Vol. 39**, pp. 1115–1193.

Glover, K. and Doyle, J.C. (1989) State-space formulae for all stabilazing controllers that satisfy an H_∞-norm bound and relations to risk sensitivity, *Systems & Control Letters*, **Vol. 11**, pp. 167–172.

Glover, K., Limebeer, D.J.N., Doyle, J.C., Kasenally, E.M. and Safonov, M.G. (1991) A characterization of all solutions to the four block general distance problem, *SIAM J. Control and Optimization*, **Vol. 29, no. 2**, pp. 283–324.

Gohberg, I., Kaashoek, M.A. and Woerdeman, H.J. (1991) Time variant extension problems of Nehari-type and band method, *H^∞- Control Theory*, Como 1990, Springer.

Green, M. and Limebeer D.J.N. (1995) *Linear Robust Control*, Prentice Hall.

Grimble, M.J. (1988) Optimal H_∞ multivariable robust controllers and relationship to LQG design problems, *Int. J. Control*, **Vol. 48**, pp. 33–58.

Halanay, A. (1990) Advances in linear control theory and Riccati equations, *Rend. Sem. Mat. Univers. Politecn. Torino*, **Vol. 48, no. 3**., pp. 251–350.

Halanay, A. and Ionescu, V. (1993) Generalized discrete-time Popov-Yakubovich theory, *Systems & Control Letters*, **Vol. 20**, pp. 1–6.

Halanay, A. and Ionescu, V. (1994) *Time-varying discrete linear systems*, Operator Theory: Advances and Applications, **68**, Birkhäuser.

Iglesias, P.A. (1991) Robust and adaptive control for discrete-time systems, Ph.D. dissertation, University of Cambridge.

Ionescu, V. and Weiss, M. (1992) On computing the stabilizing solution to the discrete-time Riccati equation, *Linear Algebra and Its Applications*, **Vol. 174**, pp. 229–238.

Ionescu, V. and Weiss, M. (1993) Continuous and discrete-time Riccati theory: a Popov function approach, *Linear Algebra and Its Applications*, **Vol. 193**, pp. 173–209.

Ionescu, V. (1995) On the evaluation of the least achievable tolerance in the disturbance feedforward problem, *Systems & Control Letters* , **Vol. 25**, pp. 113–119.

Ionescu, V. and Oara, C. (1996a) Generalized continuous-time Riccati theory, *Linear Algebra and Its Applications*, **Vol. 232**, pp. 111–131.

Ionescu, V. and Oara, C. (1996b) The four block Nehari problem: a generalized Popov-Yakubovich-type approach, *IMA Journal of Mathematical Control & Information*, **Vol. 13**, pp. 173–194.

Ionescu, V., Oara, C. and Weiss, M. (1997) General matrix pencil techniques for the solution of algebraic Riccati equations: a unified approach, *IEEE-Trans. Autom. Control*, **Vol. 42, no. 8**, pp. 1085–1097.

Iwasaki, T. and Skelton, R.E. (1994) All controllers for the general H_∞ control problem: LMI existence conditions and state space formalas, *Automatica*, **Vol. 30, no. 8**, pp. 1307–1317.

Jonckheere, E.A. and Silverman, L.M. (1983) A new set of invariants for linear systems, *IEEE-Trans. Autom. Control*, **Vol. 28**, pp. 953–962.

Jonckheere, E.A. and Juang J.C. (1987) Fast computation of achievable feedback performance in mixed-sensitivity H^∞-design, *IEEE-Trans. Autom. Control*, **Vol. 32, no. 10**, pp. 896–906.

Jonckheere, E.A., Juang, J.C. and Silverman, L.M. (1989) Spectral theory of the linear-quadratic and H^∞ problems, *Linear Algebra and Its Applications*, **Vol. 122/124**, pp. 273–300.

Juang, J.C. and Jonckheere, E.A. (1989) Data reductions in the mixed sensitivity H^∞ design problem, *IEEE-Trans. Autom. Control*, **Vol. 34, no. 8**, pp. 861–866.

Kailath, T. (1980) *Linear systems*, Prentice Hall, Englewood Cliffs, N.J.

Khargonekar, P.P., Petersen, I.R. and Rotea, M.A. (1988) H_∞ optimal control with state feedback, *IEEE-Trans. Autom. Control*, **Vol. 33**, pp. 786–788.

Kwakernaak, H. (1986) A polynomial approach to minimax frequency domain optimization of multivariable feedback systems, *Int. J. Contr.*, **Vol. 44**, pp. 117–156.

Lancaster, P. and Rodman, L. (1991) Solutions of the continuous and discrete algebraic Riccati equation: a review, *The Riccati Equation* (J.C. Willems, S. Bitanti and A. Laub, Eds.) New-York: Springer Verlag, pp. 11–52.

Limebeer, D.J.N., Green, M. and Walker, D. (1989) Discrete time H^∞ control, *Proceedings of the 28th Conference on Decision and Control, Tampa, Florida*, pp. 392–396.

McFarlane, D.C. and Glover K. (1990) *Robust control design using normalized coprime factor plant description*, Lecture Notes in Control and Information Science no.138, Springer Verlag.

McLean, D. (1991) *Automatic flight control systems*, Prentice Hall, London.

MIL-STD-1797A (1990) *Military Standard-Flying qualities of piloted airplanes.*

Nehari, Z. (1957) On bounded bilinear forms, *Ann. of Math.*, **Vol. 65**, pp. 283–324.

Nett, C.N., Jacobson, C.A. and Balas, M.J. (1984) A connection between state-space and doubly fractional representation, *IEEE-Trans. Autom. Control*, **Vol. 29, no. 9**, pp. 831–832.

Oara, C. (1995) Generalized Riccati Theory: A Popov Function Approach, Ph.D. thesis, Polytechnic University Bucharest.

Oara, C. and Van Dooren, P. (1997) An improved algorithm for the computation of structural invariants of a system pencil and related geometric aspects, *Systems & Control Letters*, **Vol. 30**, pp. 39-48.

Pappas, T., Laub, A.J. and Sandell, N.R. (1980) On the numerical solution of the discrete algebraic Riccati equation, *IEEE-Trans. Autom. Control*, **Vol. 25**, pp. 631–641.

Partington, J.R. (1988) *An introduction to Hankel operators*, Cambridge University Press.

Petersen, I.R. (1987) Disturbance attenuation and H_∞ optimization: a design method based on the algebraic Riccati equation, *IEEE-Trans. Autom. Control*, **Vol. 32**, pp. 427–429.

Popov, V.M. (1973) *Hyperstability of Control Systems*, Springer Verlag, Berlin (Romanian version 1966).

Safonov, M.G. and Verma, M.S. (1985) L_∞ sensitivity optimization and Hankel approximation, *IEEE-Trans. Autom. Control*, **Vol. 30**, pp. 279–280.

Safonov, M.G., Jonckheere, E.A., Verma, M. and Limebeer D.J.N. (1987) Synthesis of positive real multivariable feedback systems, *Internat. J. Control*, **Vol. 45**, pp. 817–842.

Safonov, M.G., Limebeer, D.J.N. and Chiang, R.Y. (1990) Simplifying the H^∞-theory via loop shifting matrix pencils and descriptor concepts, *Int. J. Control*, **Vol. 50**, pp. 2467–2488.

Sampei, M., Mita, T. and Nakamichi, M. (1990) An algebraic approach to H_∞ output feedback control problems, *Systems & Control Letters*, **Vol. 14**, pp. 13–24.

Saksena, V.R., O'Reilly, J. and Kokotovic, P.V. (1984) Singular perturbations and time-scale methods in control theory: survey 1976-1983, *Automatica*, **Vol. 20, no. 3**, pp. 273–293.

Scherer, C. (1992) H_∞-optimization without assumptions on finite or infinite zeros, *SIAM J. Control and Optimization*, **Vol. 30**, pp. 143–166.

Sparks, A. and Banda, S.S. (1993) Application of structural singular value synthesis to a fighter aircraft, *Journal of Guidance, Control and Dynamics*, **Vol. 16, no. 5**, pp. 940–947.

Stewart, G. (1973) Error perturbation bounds for subspaces associated with certain eigenvalues problems, *SIAM Review*, **Vol. 15**, pp. 727–767.

Stoica, A. (1997) Optimal H-infinity design of a control system for the angle of attack, *Proceedings of Eurocontrol, TH-AJ4, Brussels, July 1-4, 1997.*

Stoorvogel, A.A. (1990) The H^∞ control problem: a state-space approach, Ph.D. dissertation, Technical University of Eindhoven, Holland.

Stoorvogel, A.A. and Trentelman, H.L. (1990) The quadratic matrix inequality in singular H_∞ control with state feedback, *SIAM J. Control and Optimization*, **Vol. 28**, pp. 1190–1208.

Stoorvogel, A.A. (1991) Robust stabilization of systems with respect to multiplicative perturbations, *Proceedings of MTNS, Kobe, Japan.*

Van Dooren, P. (1981) A generalized eigenvalue approach for solving Riccati equations, *SIAM J. Sci. St. Comp.*, **Vol. 2, no. 2**, 1981, pp. 121–135.

Van Dooren, P. (1983) *Reducing subspaces: definitions, properties and algorithms*, Lecture Notes in Mathematics, Berlin, Springer Verlag.

Varga, A. (1992) Numerical algorithms and software tools for analysys and modeling of descriptor systems, *Proceedings of the 2nd IFAC Workshop Syst. Structure Contr., Prague, Sept. 1992*, pp. 392–395.

Verma, M. and Jonckheere, E.A. (1985) L^∞ compensation with mixed sensitivity as a broadhad mathing problem, *Systems & Control Letters*, **Vol. 4**, pp. 125–129.

Vidyasagar, M. and Kimura, H. (1986) Robust controllers for uncertain linear multivariable systems, *Automatica*, **Vol. 22, no. 1**, pp. 85–94.

Wimmer, H. (1985) Monotonicity of maximal solutions of algebraic Riccati equations, *Systems & Control Letters*, **Vol. 5**, pp. 317–319.

Willems, J.C. (1971) Least-squares stationary optimal control and the algebraic Riccati equation, *IEEE-Trans. Autom. Control*, **Vol. 21**, pp. 319–338.

Yakubovich, V.A. (1975) Continuous frequency theorem in Hilbert state and control spaces, *Siberian Math. J.*, **Vol. 26, no. 5**, pp. 1081–1152.

Youla, D.C., Jabr, H.A. and Bongiorno, J.J. (1976) Modern Wiener–Hopf design of optimal controllers: part II, *IEEE-Trans-Autom. Control*, **Vol. 21**, pp. 319–338.

Zames, G. (1966) On input-output stability of time-varying feedback systems, part 1,

IEEE-Trans. on Autom. Control, **Vol. 11**, pp. 228–238.
Zames, G. (1981) Feedback and optimal sensitivity, *IEEE-Trans. Autom. Control*, **Vol. 26**, pp. 310–320.
Zhou, K. and Khargonekar, P.P. (1988) An algebraic Riccati equation approach to H_∞ optimization, *Systems & Control Letters*, **Vol. 11**, pp. 85–92.

INDEX

Other *Mathematics and Its Applications* titles of interest:

B.S. Razumikhin: *Physical Models and Equilibrium Methods in Programming and Economics.* 1984, 368 pp. ISBN 90-277-1644-7

N.K. Bose (ed.): *Multidimensional Systems Theory. Progress, Directions and Open Problems in Multidimensional Systems.* 1985, 280 pp. ISBN 90-277-1764-8

J. Szep and F. Forgo: *Introduction to the Theory of Games.* 1985, 412 pp. ISBN 90-277-1404-5

V. Komkov: *Variational Principles of Continuum Mechanics with Engineering Applications.* Volume 1: Critical Points Theory. 1986, 398 pp. ISBN 90-277-2157-2

V. Barbu and Th. Precupanu: *Convexity and Optimization in Banach Spaces.* 1986, 416 pp. ISBN 90-277-1761-3

M. Fliess and M. Hazewinkel (eds.): *Algebraic and Geometric Methods in Nonlinear Control Theory.* 1986, 658 pp. ISBN 90-277-2286-2

P.J.M. van Laarhoven and E.H.L. Aarts: *Simulated Annealing: Theory and Applications.* 1987, 198 pp. ISBN 90-277-2513-6

B.S. Razumikhin: *Classical Principles and Optimization Problems.* 1987, 528 pp. ISBN 90-277-2605-1

S. Rolewicz: *Functional Analysis and Control Theory. Linear Systems.* 1987, 544 pp. ISBN 90-277-2186-6

V. Komkov: *Variational Principles of Continuum Mechanics with Engineering Applications.* Volume 2: Introduction to Optimal Design Theory. 1988, 288 pp. ISBN 90-277-2639-6

A.A. Pervozvanskii and V.G. Gaitsgori: *Theory of Suboptimal Decisions.* Decomposition and Aggregation. 1988, 404 pp. *out of print,* ISBN 90-277-2401-6

J. Mockus: *Bayesian Approach to Global Optimization.* Theory and Applications. 1989, 272 pp. ISBN 0-7923-0115-3

Du Dingzhu and Hu Guoding (eds.): *Combinatorics, Computing and Complexity.* 1989, 248 pp. ISBN 0-7923-0308-3

M. Iri and K. Tanabe: *Mathematical Programming.* Recent Developments and Applications. 1989, 392 pp. ISBN 0-7923-0490-X

A.T. Fomenko: *Variational Principles in Topology.* Multidimensional Minimal Surface Theory. 1990, 388 pp. ISBN 0-7923-0230-3

A.G. Butkovskiy and Yu.I. Samoilenko: *Control of Quantum-Mechanical Processes and Systems.* 1990, 246 pp. ISBN 0-7923-0689-9

A.V. Gheorghe: *Decision Processes in Dynamic Probabilistic Systems.* 1990, 372 pp. ISBN 0-7923-0544-2

A.G. Butkovskiy: *Phase Portraits of Control Dynamical Systems.* 1991, 180 pp. ISBN 0-7923-1057-8

Other *Mathematics and Its Applications* titles of interest:

A.A. Zhigljavsky: *Theory of the Global Random Search*. 1991, 360 pp.
ISBN 0-7923-1122-1

G. Ruhe: *Algorithmic Aspects of Flows in Networks*. 1991, 220 pp. ISBN 0-7923-1151-5

S. Walukuwiecz: *Integer Programming*. 1991, 196 pp. ISBN 0-7923-0726-7

M. Kisielewicz: *Differential Inclusions and Optimal Control*. 1991, 320 pp.
ISBN 0-7923-0675-9

J. Klamka: *Controllability of Dynamical Systems*. 1991, 260 pp. ISBN 0-7923-0822-0

V.N. Fomin: *Discrete Linear Control Systems*. 1991, 302 pp. ISBN 0-7923-1248-1

L. Xiao-Xin: *Absolute Stability of Nonlinear Control Systems*. 1992, 180 pp.
ISBN 0-7923-1988-5

A. Halanay and V. Rasvan: *Applications of Liapunov Methods in Stability*. 1993, 238 pp.
ISBN 0-7923-2120-0

D. den Hertog: *Interior Point Approach to Linear, Quadratic and Convex Programming*. 1994, 208 pp. ISBN 0-7923-2734-9

V.S. Tanaev, V.S. Gordon and Y.M. Shafranksy: *Scheduling Theory*. Single-Stage Systems. 1994, 380 pp. ISBN 0-7923-2853-1

V.S. Tanaev, Y.N. Sotskov and V.A. Strusevich: *Scheduling Theory*. Multi-Stage Systems. 1994, 412 pp. ISBN 0-7923-2854-X

L.D. Akulenko: *Problems and Methods of Optimal Control*. 1994, 356 pp.
ISBN 0-7923-2855-8

C. Udrişte: *Convex Functions and Optimization Methods on Riemannian Manifolds*. 1994, 348 pp. ISBN 0-7923-3002-1

B.S. Jensen: *The Dynamic Systems of Basic Economic Growth Models*. 1994, 355 pp.
ISBN 0-7923-3091-9

V. Barbu: *Mathematical Methods in Optimization of Differential Systems*. 1994, 259 pp.
ISBN 0-7923-3176-1

A. PrÇkopa: *Stochastic Programming*. 1995, 556 pp. ISBN 0-7923-3482-5

R. Lucchetti and J. Revalski (eds.): *Recent Developments in Well-Posed Variational Problems*. 1995, 266 pp. ISBN 0-7923-3576-7

A. Cheremensky and V. Fomin: *Operator Approach to Linear Control Systems*. 1995, 410 pp. ISBN 0-7923-3765-4

V.N. Afanas'ev, V.B. Kolmanovskii and V.R. Nosov: *Mathematical Theory of Control Systems Design*. 1996, 681 pp. ISBN 0-7923-3724-7

D. Pallaschke and S. Rolewicz: *Foundations of Mathematical Optimization*. Convex Analysis without Linearity. 1997, 594 pp. ISBN 0-7923-4424-3

A. Chikrii: *Conflict-Controlled Processes*. 1997, 424 pp. ISBN 0-7923-4522-3

Other *Mathematics and Its Applications* titles of interest:

I.M. Stancu-Minasian: *Fractional Programming*. Theory, Methods and Applications. 1997, 426 pp. ISBN 0-7923-4580-0

A.B. Piunovskiy: *Optimal Control of Random Sequences in Problems with Constraints*. 1997, 360 pp. ISBN 0-7923-4571-1

A.P. Abramov: *Connectedness and Necessary Conditions for an Extremum*. 1998, 212 pp. ISBN 0-7923-4910-5

A.I. Matasov: *Estimators for Uncertain Dynamic Systems*. 1998, 432 pp. ISBN 0-7923-5278-5

V.I. Elkin: *Reduction of Nonlinear Control Systems*. A Differential Geometric Approach. 1999, 260 pp. ISBN 0-7923-5623-3

N.A. Bobylev, S.V. Emel'yanov and S.K. Korovin: *Geometrical Methods in Variational Problems*. 1999, 556 pp. ISBN 0-7923-5780-9

V. Ionescu and A. Stoica: *Robust Stabilisation and H^{00} Problems*. 1999, 202 pp. ISBN 0-7923-5753-1

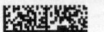